Profiting from Innovation in China

Oliver Gassmann • Angela Beckenbauer
Sascha Friesike

Profiting
from Innovation
in China

 Springer

Prof. Dr. Oliver Gassmann
Institut für Technologiemanagement
Universität St. Gallen
St. Gallen
Switzerland

Dr. Sascha Friesike
Alexander von Humboldt Institut für
Internet und Gesellschaft
Berlin
Germany

Dr. Angela Beckenbauer
St. Gallen
Switzerland

ISBN 978-3-642-30591-7   ISBN 978-3-642-30592-4 (eBook)
DOI 10.1007/978-3-642-30592-4
Springer Heidelberg Dordrecht London New York

Printed on acid-free paper

Springer is part of Springer Science+Business Media (www.springer.com)

Contents

Introduction:
Innovation in China—Paradox or Paradigm?

The basic mechanism of Chinese competitiveness is simple: low wages lead to low production costs, low prices in niche markets lead to higher volumes, small businesses turn into large businesses which in turn leads to even lower costs. Parallel to this self-enforcing cost reduction spiral, competitors in emerging markets build up knowledge and experience. They evolve from imitators to innovators. At first, innovation happens rather slow, but speed significantly increases over time.

Today the Chinese share of the world market is considerable: 40 % of all TVs, more than 50 % of all digital cameras, 40 % of all mobile phones, 60 % of all toys, and 70 % of all marine containers are made in China. The share of those products not only manufactured but also designed in China is rapidly growing, too. China is dramatically catching up and is on the verge of becoming a technological innovator. The number of Chinese firms with global ambitions is growing rapidly. More and more technological innovations are coming from China, and the number of patents in China is growing steadily.

"The rise of China" we constantly hear about in the media is hardly a rise but merely a comeback. We have to go back only a couple of hundred years to see that China once was one the most sophisticated civilizations of its time. When what we today consider 'the West' was basically barbarians fighting in the woods, China was a promising and flourishing society. The Chinese invented paper (200 a.c.), bone china (300 a.c.), the magnetic compass (300 a.c.), the letter-press (750 a.c.), and of course gun powder (1000 a.c.). China's rich history also explains its current national identity as China is not an aspiring spark but rather an old-established civilization that simply wants to be back to where it was only centuries ago. Innovative firms such as Huawei, Lenovo and Haiers are therefore logical consequences.

This might remind the world of the Toyotas or Samsungs from decades ago. But the Chinese growth is expected to be faster than the Japanese revolution in the 80ies or the Korean revolution in the late 90ies as Zheng and Williamson wrote in 2007. Today the globalization level is higher, which speeds up the entire process. Modular supply chains, higher levels of outsourcing, massive foreign direct investments, and the gigantic home market will enable the Chinese firms to establish a much faster industrialization (Fig. 1).

'Innovation' is not just an abstract concept to the Chinese. It can rather be felt (see ch-ina.com): "In recent months, when taking the new high speed trains out of Shanghai, a time-conscious frequent traveler would have noticed that his train left

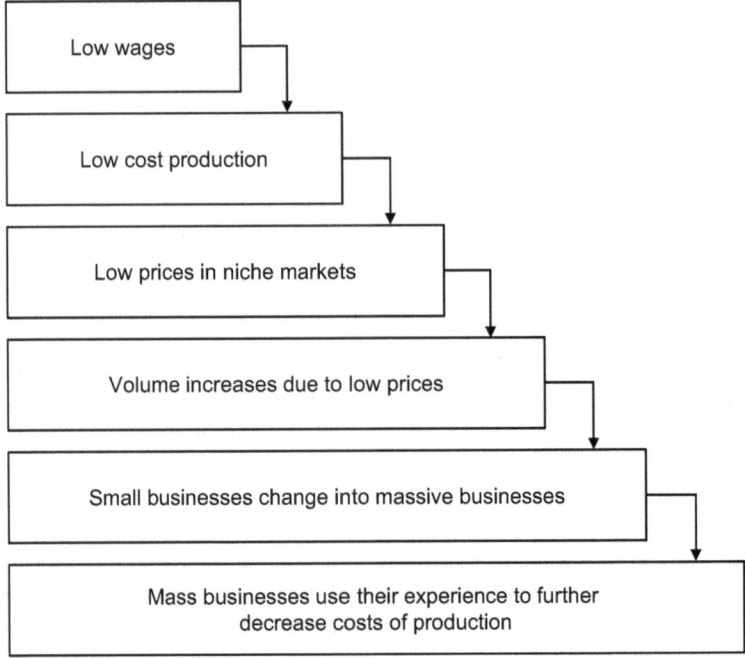

Fig. 1 Mechanism of Chinese competitiveness

the station at least 1 minute early, every time. It needs careful attention to notice because when the train leaves there is no commotion of last minute passengers trying to get on: the platform is empty and every traveler is already on the train, early as well."

It feels like the whole country is getting on board of China's advanced innovation schedule. 42 high-speed train lines (13,000 km of track with a high-speed of 350 km/h) are under construction. Initially, the lines were planned for completion in 10 years time. But when the world financial crisis hit, investments were made by the Chinese government and the project timelines were brought forward to finish all lines by 2012!

An army of more than 100,000 workers build the 1,500 km from Shanghai to Beijing. This line will cut rail transportation time from 12 h down to less than five—making the train trip competitive with flying while saving considerable amounts of fuel. Just as China's transportation sector is accelerating so is every other facet of the Chinese civilization. China plans to build three times more nuclear power plants than the rest of the whole world combined (approximately one new nuclear plant every two months) by 2020. That is in addition to new coal power plants—a new coal power plant is going online every single week.

BYD, which in 1995 was a battery producer with only 20 workers, today employs 130,000 people and produces hybrid cars with exclusive battery technology. The company received a USD 250 million investment from Warren Buffet's Berkshire Hathaway, and signed a deal with Daimler Benz to co-produce a Smart e-car for China's domestic market. The Chairman of BYD—BYD stands for 'Build Your Dream'—intends to live by his company's name. He plans to become the world's biggest car manufacturer. Being only 47 years of age, he might very well succeed considering the push that China is making towards electronic transportation. 13 cities have already been earmarked for the complete electric conversion of public transport (and that includes taxis). Incentives for private buyers of electric vehicles are in the making.

Every year over 6,000 PhD students receive a scholarship to study abroad in exchange for the commitment to return to China afterwards. They work in public research labs and universities or they build their own start-ups based on their technological inventions. In terms of research China is making a staggering progress: quantitatively it is one of the largest source of scientific publications already. In selected fields, the Chinese clearly aim for a global leadership position in a few years from now. In the material sciences for instance, China publishes more than 20 % of the world's scientific articles—and more than 30 % in the sub-fields of metallurgy and crystallography (ch-ina.com).

Chinese firms have an accelerated way of decision making if they really want to. At the expense of neglected corporate governance managers are empowered and not restricted by the western shareholder mentality.

But it is not just sunshine and roses in China. There are a lot of national challenges such as huge regional and social differences, an unstable banking system, corrupt companies and institutions, ecological challenges due to the fast economic growth, and high energy consumption. Innovation and the protection of intellectual property are important goals for the government, but not the only ones. The political stability is one of the most important drivers for innovation. Only if these big social, ecological and economical challenges will be met, the Chinese innovation system will further boom and prosper.

One of the biggest barriers for western firms shifting technology intensive processes or research and development to China, is the lack of enforcement of intellectual property rights. China has a sophisticated patent system, which has been developed in accordance with the German system. But the enforcement of IP rights in one of the 3,000 courts all over the country is extremely difficult. Today the registered IP violations in China (2004: 23,500) are only the tip of the iceberg. Because of the low probability to win the slow process of enforcement, most of the firms prefer to arrange themselves with the pirates within China and sue them at the border to Europe or the US. Or they just try to be faster and better.

"China is the market of the future—and it always will be ..." told us a CEO of a large machinery firm. Doing business in China is not easy—and managing intellectual property rights in China is one of the biggest challenges. If firms do not succeed in protecting their innovation and new technologies from imitators, they give away their temporary monopolistic profits. This competitive advantage has been the reason why they have been invested in research, development and innovation.

Even though China is becoming an evermore-important market to any internationally active firm, profiting from innovation within this market is connected with several challenges. In this book we present how firms can profit from innovation in China nevertheless. This book is for managers who are doing business in China as well as scholars and students of International Management and of Technology and Innovation Management.

The groundwork of this book is based on several years of research with dozens of companies, which have either shifted their R&D or technology intensive processes to China or which suffered from being imitated by companies from Shanghai, Guandong or somewhere in the big Northwest of China.

Our book is divided into five chapters. The first chapter lays the foundation and explains why innovation and China go indeed hand in hand. The second chapter explains why the main difference between the Chinese market and western markets lies with intellectual property. The third chapter introduces the dimensions of this difference. The fourth chapter introduces the 'IP protection star'—a five dimensional framework, which explains the major challenges when dealing with IP in China. The fifth chapter presents an integrated management model. The sixth chapter presents hands-on managerial implications. And lastly, the seventh chapter presents an outlook.

Profiting from innovation in an emerging market such as China is a major challenge to any company. We hope that the tools and practices we present in this book will be helpful to those engaged in China. We have learned the presented practices by working with and researching some of the most professional firms in the world—we are confident that the content will be helpful to many.

April 2012   Oliver Gassmann
St. Gallen, Schaan, Berlin   Angela Beckenbauer
                                                  Sascha Friesike

China as a Place to Innovate

Many western firms have to find an answer to the massive and aggressive Chinese growth. In the past it has been relatively easy: China was a destination for low cost manufacturing and a market for low cost products. Today this is different: China has become very competitive in many technological areas. This is also due to the successful agreement many if not most Chinese joint ventures work with: "You give us your technology and we provide you with access to our huge market."

'Technology for market' worked in most cases—at least for the Chinese economy. This market deal has been complemented by an extensive buy-in of foreign experts. These western consultants, engineers and scientists have been very expensive compared to local fees but very effective, since they accumulate the technology and the know-how of a few hundred years of experience (Zeng and Williamson 2007).

1.1 China Is the Number One Place for New R&D Labs

The continued growth of the Chinese economy accompanied with the expansion of international investment in China has led to an increase of foreign research and development (R&D) activities in the country. Aside from the rising importance of R&D internationalization, research on foreign R&D in China has been neglected in the past due to its nascent state.

More and more transnational companies have identified China as a preferred place to conduct offshore R&D activities. Ten years ago almost all R&D centers were based in the two most economically important cities of China, Beijing and Shanghai. Today, the two cities are still playing a dominant role but many other Chinese regions are seeing foreign R&D investments, too. The education level in China is constantly rising and the infrastructure is ever improving. These are the two basic reasons why many regions besides Beijing and Shanghai are becoming R&D centers.

The computer and telecommunications industries drive R&D investments in China. But they do not stand by themselves. Other important industrial branches

O. Gassmann et al., *Profiting from Innovation in China*,
DOI 10.1007/978-3-642-30592-4_1, © Springer-Verlag Berlin Heidelberg 2012

with R&D investment in China include chemical, petrochemicals, biotech, pharmaceutical, automotive, transportation and power generation equipment. These multinational companies typically come from the triad regions. Most of them are from North America, especially from the U.S., followed by the European Union and Japan. A further significant group of R&D investors come from Greater China, and Taiwan.

The strongest driver for foreign R&D has always been the market. If the company's business requires local product adaptation and intensive customer cooperation, it is likely that local development units will be established (von Zedtwitz and Gassmann 2002). But today we see many firms that go far beyond innovating for China. Their Chinese R&D labs are responsible for many of their breakthrough-innovations in recent years.

1.1.1   Motivations for Establishing R&D Labs in China

Prior to studying western companies' motivations for establishing R&D in China, we shall briefly outline the general drivers for conducting R&D abroad. Different approaches have been applied to classify motivations for R&D internationalization. One approach broadly distinguishes between demand-oriented and supply-oriented drivers for the internationalization of R&D (see Granstrand et al. 1993; Dunning and Narula 1995; OECD 1998; Doz and Hamel 1998). *Demand-oriented* motivation factors include the special needs of the local country/market, which require modifications of firm's products; or host country restrictions, such as local content requirements, tolls, import quotas, and fulfillment of standards. *Supply-oriented* factors include highly sophisticated foreign scientific infrastructure (e.g., new regional technological competence centers such as Silicon Valley, Prato or Modena), which takes advantage of host country scientific and knowledge inputs and accesses cutting-edge technology. Availability of well-educated local R&D specialists, ideally combined with low R&D personnel costs are further supply-oriented incentives to establish R&D abroad. A third group of motivations, environmental motivation factors, is mentioned by Granstrand et al. (1992).

In a more refined classification scheme, Beckmann and Fischer (1994) identify five categories of R&D internationalization drivers (input-oriented, output-oriented, external, efficiency-oriented and political/social-cultural). This classification has been used to classify drivers of R&D internationalization whereas the input- and output-oriented factors are principally in accordance with the supply- and demand-oriented view. The three other categories such as external, efficiency-oriented, and political/social-cultural factors reveal the multifarious character of motivations for R&D internationalization.

Based on our research interviews and literature analysis, we have relied on this scheme as a preliminary framework to examine the specific motivations for establishing R&D in the Chinese context. We have merged output and efficiency-oriented motivation factors into performance-oriented motivation factors and the

Table 1.1 Important motivations for western companies' R&D establishment in China

Input-oriented motivations	Performance-oriented motivations
Availability of high qualified personnel	Customer and market-specific development
Tapping informal networks and knowledge source	Adaptation to local production processes
Local pocket-of-innovation	Cost advantages
Business-ecological motivations	
Governmental policy	
Continuing economic growth and unique market size	
Peer pressure	

external and political/social-cultural into business-ecological motivation factors (see Table 1.1).

1.1.1.1 Input-Oriented Motivations

Availability of Qualified Personnel Amongst the input-oriented motivations for establishing R&D in China, the huge human resource potential is of great importance. In the last 10 years the Chinese education budget has increased by 20 % (9 % since 2009 alone). Western countries were rolling out massive budget cuts in the same period. Currently, there are 1,550 Universities in China, 725,000 professors and 11 million students. Annually 2.5 million students graduated from the countries' universities including 14,000 Ph.D.s, ranking China third behind the U.S. (approx. 40,000) and Germany (approx. 30,000, Ministry of Education P.R.C. 2003a). Many top universities such as Tsinghua, Beijing, Zhejiang and Fudan train highly qualified students in disciplines such as mathematics and natural sciences.

Currently, more than 700,000 students are studying abroad. The majority of students chose to emigrate to other countries after graduating. Hence, China has suffered from an outflow of talents (brain drain) to a great extent. In recent years, Chinese governments, at both national and local levels, have introduced policies to induce highly skilled overseas Chinese to return to China. Increasing numbers of scientists and graduates have returned from abroad thanks to the enduring economic growth and better opportunities in China. A famous example is the company 'Wuxi Apptec', which was solely formed by returnees and has since become a dominant player in the pharmaceutical world. In only 10 years, Wuxi has grown to more than 4,000 employees.

Tapping Informal Networks and Information Sources In China, business success is heavily dependent on good informal networks and relationships—the frequently cited 'guanxi'. The establishment of a local R&D center enables a company to build and maintain informal networks with universities and local scientific communities, which can help western companies to establish strategic partnerships and establish human resources on a long term basis. In addition, China's industrial development is at an emerging stage and the economy is undergoing a transition from a planned to a market based system. Hence changes in industrial regulations, legislations and policies are all the more dynamic. Their on-spot R&D activities and proximity to the government help western companies to keep pace with changes in the dynamic Chinese environment and allows them to achieve critical competitive advantages.

Using local R&D to gather technology intelligence on local and international competition is yet another input-oriented motivation for western companies to conduct R&D activities in China.

Local Pocket-of-Innovation Since Chinese policy makers seek to raise the level of China's industrial production and increase the country's competitiveness to an international level, special economic and other investment zones have been established and in doing so have become the main engine for growth in the Chinese economy. Notably, the High Technology Development Zones (HTDZ) or 'science parks' have been designed to lure researchers, entrepreneurs, foreign R&D centers and venture capitalists from around the world.

As an example, Beijing's high-tech Zhongguancun area located northwest of the city is home to a large number of universities and scientific institutions including Tsinghua University, Beijing University and the Chinese Academy of Science. As a result, there are a number of start-up firms, foreign-capital firms and large-scale local firms that are seeking access to high potentials through building strong relationships with the universities. Furthermore, these pockets-of-innovation attract investors with space, advanced infrastructure, and high-tech facilities they require along with financial incentives. As an example, the Chinese State Council and Beijing municipality both offer start-up firms located in Zhongguancun area tax-free operation for 3 years following their establishment, followed by a 50 % discount for the next 3 years, and a 15 % discount from the seventh year onwards, along with other tax incentives. Due to the substantial governmental support and geographical uniqueness, it is not surprising that several interview partners believe a few of these industrial and science parks will become centers of excellence in the future. Since more and more Chinese cities and regions are trying to capture the attention of western companies by various incentives, more western companies have invested outside of Beijing and Shanghai, the two established hubs of foreign R&D activities in China.

1.1.1.2   Performance-Oriented Motivations

Customer and Market-specific Development One main reason why so many companies establish development bases in China is to locally develop products specifically for the Chinese market. The necessity of adapting products to the foreign market is a widely shared belief of many interviewed R&D managers. Selling products without paying attention to the needs of the local markets is bound to fail. Locating R&D activities in China allows western companies to adapt and tailor their products and services to the local culture and market needs. A typical example is adapting IT user interfaces, telecommunication or car infotainment products to be used via the Chinese language. Moreover, specific local conditions in which products are operating require appropriate modification and redevelopment. For example, in China some automotive components such as air conditioning and combustion engines need to be redeveloped according to local climatic conditions and local gas quality. Within the next 2–3 years over 75 % of growth in electronic manufacturing capacities will take place in China. Risks of such a production shift purely for cost reasons are high; local development and product adaptation in these fast growing markets can support manufacturing operations and increase competitiveness.

Yet, many firms go even further. The electronics company 'Philips' for instance sees China as their second home market. There is an additional value to operating in different cultures and countries such as China. A company can develop new products and forge advanced thinking on many product issues. Some managers do believe that products, which satisfy the requirements of the most difficult consumer and market environments are likely to succeed anywhere in the world. The Microsoft Research Center in China pursues problems of computing in Chinese due to the difficulty of inserting Chinese characters onto a western keyboard. Besides the improvement of software's suggestion and error-checking systems, researchers also focus on data entry methods such as speech and handwriting recognition. The result will make computers more user-friendly in Chinese, but will in the end benefit all computer users.

The elevator and escalator company Schindler established an R&D center in Shanghai in the late nineties because Shanghai is one of the most booming and sophisticated markets in the construction business. Chinese customers are less risk-averse than western customers, which is typical for booming economies. In 2003 Schindler conducted a field study for a new web-based personalized infotainment system in the elevator cabin—an advanced experiment, which would be less likely to be accepted in Europe or the US. Based on that study Schindler has planned to multiply the system requirements for the product launch worldwide.

Cost Advantages Cost advantages have long been a main motivation for R&D centers in China. Yet, experience shows that the cost advantages are by no means as extensive as they were predicted only 10 years ago. Labor costs (especially for the highly educated) have increased dramatically in recent years. Still, R&D personnel in the U.S. or in Europe is far more expensive, yet the overall cost advantages are not that impressive—when one takes coordination costs into consideration.

Short R&D Cycle Time and Adaptation to Local Production Processes Localized R&D allows for a shorter R&D cycle time especially for products, which require customer and market-specific accommodations. Furthermore, local R&D activities can assist manufacture operations to improve quality, learn to produce new offerings, reduce costs, or increase capabilities.

A Different Take on Intellectual Property

When you stroll around China's enormous metropolitan areas it is challenging not to notice a take on intellectual property that is fundamentally different from the one in the West. You can see street vendors claiming to sell "original" phone accessories or "genuine" Swiss watches. Logos of famous brands are often used to convey the idea of quality and resemble something famous. The picture below shows a convenience store in downtown Shanghai. The store's elaborate lighting resembles that of the US fast food chain McDonald's. Even the smaller sign below the logo "24open" is a re-miniscence of the burger chain.

This convenience store is not an exception. Stores that resemble famous brands are quite typical in metropolitan China. There are coffee houses that look a lot like Starbucks (they are named Seayahi Coffee, Lucky Coffee or Buckstar Coffee), fast food vendors that are named KLG with a logo that looks like the one from KFC, Nibe stores showing the famous Nike swoosh, among many others.

The city of Beijing even runs an amusement park called the "Beijing Shijingshan Amusement Park" that looks a lot like Walt Disney's iconic Magic Kingdom.

1.1.1.3   Business-Ecological Motivations

Governmental Policy For almost two decades now, one can observe China's increasing sensitivity towards technology's contribution to economic growth. 'Revitalizing the Nation through Science and Education', a strategy which was officially adopted in 1995 by the Chinese government to speed up scientific and technological progress, and has led to rapid growth of China's national science and technology activities. In 2006 China's R&D intensity reached 1.43 % (up from 0.6 % in 1995). R&D spending has increased at a remarkable annual rate of 19 %. Since 2000 China has been ranking second in the world in terms of total number of researchers (only behind the U.S. and ahead of Japan). Today, China has the ability to attract long-term, relatively capital-intensive and high-tech projects from multinational enterprises in OECD countries (OECD 2008). As a result, China has continued to liberalize the approval process for FDI and a number of preferential policies have been put in place in order to encourage foreign business, especially western companies, to set up local R&D investments.

Chinese policy makers believe that an effective way to bridge the gap to the international technology level is to intensify the linkage to the international R&D community. One important means is the establishment of high-tech parks combined with incentives, such as free rent, low tenancy costs, favorable lease terms, and tax relief.

As identified by Ambrecht (2002), there are several multi-faceted reasons behind these kinds of policies. Foreign laboratories will bring capital investment, ancillary spending, and job opportunities. Moreover, they help to attract excellent ethnic Chinese specialists from around the world back to China to conduct advanced research. The proximity to international research facilities will also spur the Chinese high education system through their demand of local high-quality technical personnel and co-operation with Chinese R&D facilities.

Moreover, the business background of these R&D laboratories could help China create market value out from the leading-edge technologies being developed in Chinese universities and research institutions. Furthermore, local R&D activities are considered to be important evidence that the company is interested in developing long-term commitments in China. It helps to build trust and good working relations with the government and to receive official support. But due to the enticement of financial incentives and other business advantages, some foreign firms are even tempted to register their China activities as 'R&D', whether their research does or does not entail genuine research and development activities (see Walsh 2003).

Given the pure power of the Chinese government, they are in a position to play one foreign investor against another in order to accelerate western companies' investment level and R&D commitment. Prior to China's accession to the WTO, foreign investors were regularly pressured to transfer technology in return for market access.

Continuing Economic Growth and Unique Market Size Aside from stagnated world economy, the dynamic growth of Chinese national economy and its overwhelming market size has made China amongst the most strategically important markets for western companies. Especially in IT and telecommunications, multinational giants such as Microsoft, Nokia, Apple, and Siemens have invested hundred of millions of dollars into their R&D activities in China, which is in essence an investment in China's future market. For example, China is the world's largest mobile phone market with more than 900 million users by the end of 2009 (census. gov, 2011). The critical mass of the Chinese and the Asian markets is increasingly influencing mobile phone size, style and applications globally. Traditionally China was a market for cost-effective entry-level phones. But in recent years it has also become a main market for margin-rich smart-phones.

Peer Pressure Western companies' motivations for establishing local R&D in China are rooted in the awareness of possible mid- and long-term competitive advantages, which have been discussed in the above sections. Seeing that western companies' competition on the Chinese market has intensified and the number of foreign-invested R&D centers in China has grown in recent years, those who do not have such investments have come under increasing pressure to invest in R&D. Even though most interview partners did not want to admit to peer pressure as a driver, it had been mentioned during informal follow-ups.

1.1.2   Barriers for Managing R&D in China

As discussed in the previous section, China is a very attractive location for transnational companies' R&D units. On the other hand there are still high barriers before exploiting that potential. Despite the aforementioned advantages and rewards for setting up R&D activities in China, the following barriers could neutralize them to a certain extent.

Difficulty in Management Due to Language and Culture   Given the general lack of experienced indigenous R&D managers in China, the majority of upper R&D management is staffed by foreign expatriates. Unfortunately, most of them do not have adequate or no management experience in the Chinese environment. The Chinese language is an initial barrier in management. Although some of the top Chinese research staff has a good command of English, most of the local engineers only have limited English language capabilities.

An even greater obstacle for western managers is to overcome the cultural gap during the daily interactions concerning issues like communication style, 'face saving' etc. A western manager may have done everything correctly according to their understanding of good management style. However, the lack of experience and sensitivity to Chinese mentality and culture will usually incur managerial inefficiency, wrong decisions and inadequate leadership. Western managers coming mostly from low context cultures (e.g., German, U.S.), are used to capturing the meaning of a message with words alone. They believe spelling it out clearly is the only way to avoid ambiguity. On the contrary, the Chinese culture is a very high context culture (Hofstede 1994). A message is delivered with nonverbal signals (e.g., tone of voice, use of silence, facial cues, and body language), unspoken assumptions, and the context or environment surrounding the conversation. People from high context culture assume that the receiver of the message is intelligent enough to understand its true connotation. Lack of awareness and proper handling of interference between high and low context communication styles can eventually lead to misunderstanding, confusion and ineffectiveness.

Diversity of R&D Staff The R&D teams of western companies in China are diversified and typically composed of three groups of people. Local graduates make up the majority of the R&D staff. Western expatriates and global Chinese comprise the other two groups of the team. Although diversity in R&D teams can increase creativity and innovation, it also provides sources of potential conflict. In addition to general difficulties of managing intercultural teams, one particular challenge lies in the potential conflicts between the local Chinese staff and the global Chinese.

In our context, 'global Chinese' is a generic term for three subgroups of Chinese people working for foreign R&D: Mainland Chinese returnees with education and working experiences abroad; Chinese from Greater China (i.e., Taiwan, Macao, and Hong Kong); and overseas-born ethnic Chinese. On the one hand they share the same Chinese origin and culture and have almost no language difficulties with each other. On the other hand due to multi-layered differences between these subgroups, due to such elements as different educational backgrounds, different working styles and perceptions, and in particular the huge gap in pay for the various levels (see also De Boer et al. 1998), one should be wary to generalize these three sub-groups of Chinese people. Western expatriates are often not aware or underestimate these differences.

Low Individual Initiative and Innovative Mindset The majority of local R&D staff is recruited from China's leading universities. During the interviews, most of the managers shared the opinion that Chinese graduates have a solid education and are highly skilled in solving certain well-defined tasks. But there is an awareness of a lack of practical experience and individual initiative, which is to a degree in line with the findings of Walsh (2003, p. 96). It could be argued that this phenomenon is attributable to the Chinese education system, which is characterized by a narrow curriculum design and a neglect on the development of individual initiative.

To a greater extent, an R&D staff member's individual initiative is decisive for creativity and innovation. As Walsh (2003) stated, developing a more innovative mindset among Chinese staff is a primary concern of foreign R&D managers at this stage. Risk taking behavior and entrepreneurship in the widest sense have to be promoted. As a result, the management and the development of R&D staff in China require much attention.

High Employee Turnover Rates and Lack of Loyalty Like many other foreign enterprises in China, many R&D departments are plagued by high staff turnover rates, particularly those located in large cities such as Beijing and Shanghai, where sufficient new opportunities are available. In general, there are three main causes of staff turnover. Some of them who leave because they have simply found a better paid job elsewhere, while some go abroad to obtain graduate degrees. Only a few, but worth mentioning, leave to work for or establish high-tech start-up enterprises. This is a common phenomenon not only impacting foreign companies in China, but Chinese domestic high-tech companies and research institutes are suffering from high turnover as well. Foreign companies are often used as a career springboard. Working for a foreign company provides Chinese graduates not only with a higher salary and practice experiences; it also makes them familiar with western management practices and provides them with possibilities of advanced on-the-job training. These kinds of references enable them to get a job with better payment and perspectives. In the worst-case scenario, competitors would hire them.

As stated by several foreign managers and also confirmed by Chinese employees, compensation strongly influences the affiliation and loyalty of Chinese R&D staff. Beside money issues, one should not ignore the level of attachment to western employers. It could be argued that given the fact that China has made a strong effort in building national consciousness; many Chinese employees strongly associate themselves to their own country than to their western employer. As a result some R&D managers have expressed their lack of trust in local people, since they are afraid of lack of loyalty and loosing knowledge to the competitor. This is especially relevant within Sino-foreign joint ventures whereas the parent company of the Chinese partner or simply the partner themselves participates in or owns other domestic enterprises operating on similar business fields. For example, ABB has lost plenty of technological knowledge through their Swiss-Chinese joint ventures. The lack of trust in local people is one reason why Schindler's competitive intelligence unit in Switzerland consists of two Chinese staff members.

Building long-term staff loyalty is a challenge for human resource managers in China. It is particularly relevant for R&D labs, given that know-how travels with people.

Bureaucracy and Uncertainty in Legal Changes As mentioned in the above section, the Chinese government provides incentives for foreign R&D activities in China. According to the experiences of some interviewed R&D managers, receiving promised preferential conditions such as tax relief and other incentives can be a stressful and prolonged procedure, due to multiple bureaucratic hurdles and very specific rules. Importing test materials can be difficult (IBM), transferring people from Beijing to Shanghai requires an official permit (requires long-term preparation) (Siemens, ABB).

Therefore, a good relationship ('guanxi') network with the government is crucial to business efficiency and success. This kind of relationship network needs time and occasional financial support. As one western expatriate manager mentioned, relationship investments takes the form of sponsoring of IT equipment for local universities or other contributions to non-profit official organizations such as municipal kindergartens. However, one should not mistake this kind of financial aid for a bribe.

Due to a lack of transparency in Chinese policymaking, China's industrial, political, legal, technological policies and strategies are difficult to discern. This provides more uncertainty for foreign R&D activities in China. Furthermore, even if the intervention on foreign enterprises' activities by the Communist Party of China (CPC) has decreased in recent years and the party branch (i.e., the party secretary) within some wholly owned foreign company is not involved in the business at all, some interviewees still mentioned that the a strong governmental influence remains. As one manager said: "There are still numerous possibilities for (the) Chinese government to make everything difficult."

Case Example Stihl

A Stihl importer reported the brand name '7HILS' on a chain saw packaging. He found that the font, letter style and color looked similar to the Stihl logo. By turning the logo upside down and mirroring it, the 7HILS brand name looks very similar to the Stihl brand.

Trademark *OLO-7HILS* of Chinese competitor

Trademark turned upside down and mirrored

Original *Stihl* trademark (e.g. used on *Stihl Chain Saw 070* Model)

A few years after the copy of the older model, an imitation of the new cut-off machine model TS 400 emerged on the market. At that time, the product was trademark- as well as patent-protected in main markets such as Germany, France, Great Britain, Japan, the USA and China. The copy clearly infringed the patent-protected air filter and the starter on the right-hand side as well as the semi-automatic belt tension. In addition, the design was similar to the Stihl product with a very similar color scheme; it should be noted that the Stihl trademark was not applied on the product. This incident was a turning point in the IP application strategy. Since then, Stihl consistently applies its relevant patents in China. So far, the only entire copy of a Stihl product to have been discovered was in China, and was of Chinese origin—that of the TS 400 cut-off machine.

Original TS 400

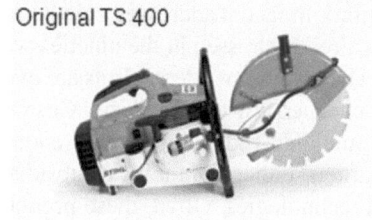

Copy

1.2   Siemens' SMART Principle and Reverse Innovation

Siemens is facing the Chinese challenge in nearly all business areas and has developed a corporate strategy to meet this challenge. The answer of Siemens is based on the SMART principle, which has the following cornerstones:

S   imple
M   aintenance friendly
A   ffordable
R   eliable
T   imely to market

The SMART principle addresses a very western characteristic: Over-engineering and technology-driven innovation without a clear customer value. Overall, about 30 % of all innovations flop due to over-engineering, also known as 'the electronic mouse trap'. This describes an innovation that is based on a highly sophisticated technology, loved only by the developing engineers. The customers however, regard it quite differently and often dislike the product's complexity and its price-point. The importance of delivering customer value cannot be stressed enough.

If it succeeds the innovation is not only competitive on the home market but on the world market, too. In a benchmark-survey on western R&D in China conducted by the Institute of Technology Management in St. Gallen (2006) we discovered an interesting phenomenon: Most western countries shifted their R&D to China because of the large Chinese market. Hence, the motive was to exploit the Chinese market. But 75 % of all firms exported the Chinese-invented products to the world market later on. What is good for the Chinese market can be good for the world market, too. What works for the Chinese market also works for the world market.

This phenomenon is called 'reverse innovation'. In fact, it addresses a challenge, which Harvard colleague Clayton Christenson already stressed in the nineties with his seminal work 'The Innovator's Dilemma': The leading (western) firms are over-engineering while competitors come up with cheaper low end products. Customers consider them not as lousy but as 'good enough'. Today, these good enough products are developed for local, emerging markets, either from a local subsidiary or from a local competitor based on outdated technologies. Often, these products cater to the general need of a more cost effective alternative to the products available on western markets. To answer this need companies start to export and make the cost effective 'good enough product' available on western markets. The process is called 'reverse innovation'. As such, reverse innovation describes a trend where innovations are developed in emerging countries (China, India, Brazil …) and are later commercialized on western markets. A famous and often cited example is the

Chinese appliances firm 'Haier'. Haier developed a small washing machine for the everyday use. The machine was later successfully commercialized in both, Europe and the United States. Another example comes from GE. Their Shanghai R&D subsidiary developed a low-cost portable ultrasound machine for the use in rural China. The machine was later a big success in United States' hospitals. Further evidence for this trend can be found in literature (Immelt et al. 2009; Zeschky et al. 2010).

It is interesting to see why so many innovations developed in emerging countries turn out to be successful in the west, too. Studying the many known cases similarities among them can be found. The products are usually simple in their design and are therefore easy to repair. They are low-cost products but at the same time no low-quality products. Furthermore, they address simple needs in a precise and cost-effective manner. Most products invented in China have one big competitive advantage: they are very cost competitive. In the cement industry, firms like the world market leader Holcim accepts different design criteria of the Chinese suppliers when the price is 35–50 % lower than that of the western competitors. Within only 5 years Chinese suppliers of Holcim gained a market share from zero to 30 % in 2010, mostly by offering a very cost-competitive solution. In most industries western machinery firms counteract this Chinese challenge with higher quality.

The underlying sociological reasons are quite intelligible, too. Only a decade ago western firms, active in emerging markets, aimed at the top of the population pyramid. Emerging markets were basically the few rich people within these markets. The mass was simply neglected. Only when new innovations were targeted at the millions of people (too poor to buy western high quality products but still in need of let us say a washing machine) did the phenomenon of reverse innovation occur.

The need for simple, reliable and affordable products is not limited to emerging markets. In the west this need was further strengthened by recent recessions. Having less money to spend more and more consumers are looking for reliable, yet affordable products. Increasing labor costs also added to the equation, as complex and sophisticated products go along with high maintenance costs. In our western societies, in which people own more things than ever before in history, maintenance costs for every single piece of equipment rapidly add up.

Checklist for Managers of R&D Operations in China
- How to manage transnational R&D projects in China?
- How to implement milestones and controlling of the local projects?
- How to manage the interfaces to the other R&D labs?
- How to maintain the quality level?
- How to integrate the local operations into a global platform strategy?
- How to implement concurrent engineering practices to China?
- Is the infrastructure, e.g., ICT, good enough?
- How to get local customer requirements?
- How to integrate local customers into the innovation process in order to activate the local user knowledge?

- How to leverage central knowledge into the Chinese operations?
- How to keep and enforce the corporate standards worldwide?
- How to design for local manufacturing?
- How to design for local supply chains?
- How to design for local suppliers with lower tolerances in production?
- How to integrate the Chinese operations into the global system?
- How to promote global technology transfer and organizational learning in the local environment?
- How to attract and moreover keep talents?
- How to attract Chinese employees with bi-cultural backgrounds?
- How to design the incentive system for the engineers and managers?
- How to cooperate with Chinese partners?
- How to protect the IP in China?
- How to increase the awareness and sensibility of local engineers for intellectual property protection?
- How to get access to the local universities?
- How to develop 'guanxi' for relationships with stakeholders, which can be used later on?

In order to stay successful with R&D and technology intensive operations managers have to:

1. Understand the Chinese culture/at least develop empathy for the differences.
2. Develop human linkages: The right person in the right place at the right time bridges cultural differences.
3. Adapt the management style to the local mentality. What works in Europe and the U.S. does not necessarily work in China, too.
4. Use expatriates in the first 3–5 years for implementing the corporate virus, typically they need to stay longer.
5. Be careful in your choice of local partners; do not expect loyalty.
6. Get access to the local universities; they are the matchmakers and gatekeepers to local partners.

1.2.1 How to Enable Reverse Innovation?

All of these trends put together can explain both, the success of reverse innovation and the intense interest many corporations show towards it. In studying the phenomenon we found that firms have to successfully undertake several steps in order to enable reverse innovation:

Increased Autonomy of the R&D Subsidiary Many firms still treat their R&D subsidiaries as a sort of extended workbench. Allowing them only to adapt the company's products to the local market. However, these R&D subsidiaries are not allowed to develop their own products let alone to develop products the western firm does not see a market for. Yet, only when the autonomy of the R&D subsidiary is increased and corporate design rules are put on hold, is there a chance for reverse innovation.

The Swiss elevator and escalator company Schindler first shifted parts of its R&D to Sao Paulo and later to India. The main reasons for the two relocations were the local design-to-cost capabilities. Setting up an R&D lab in Shanghai followed the same rationale: Chinese engineers earn less (which is often compensated by an increase of other costs related to global R&D), but Chinese engineers do not suffer from the 'over-engineering-gene'—a popular disease among German and Swiss engineers. They design simple, robust, and affordable products. Yet, lower quality products imply the risk of high warranty costs; they can also harm the company's brand. Empowerments of subsidiaries are fashionable, but opportunities and risks go hand in hand.

Local Reverse Innovation Capabilities Local teams have to learn product development from scratch. Nowadays, many R&D subsidiaries are limited to enhancing products given to them. This however is a completely different task to developing an entire product. Innovation management tools for all phases of the product development have to be learned by the local teams in order to successfully develop. These phases include everything from 'idea generation' to 'need finding' to 'prototyping' to 'testing' and eventually 'launching'.

Modern Internet based communication and information instruments such as crowdsourcing can speed up the learning process in the local labs drastically. Jam sessions as IBM's famous 'Innovation Jam' offer firm-wide opportunities to participate in the corporate innovation process.

Allowing an East-to-West Flow of Innovation It is necessary to rethink the idea of innovation flow. Today, most companies do not appreciate the innovation efforts undertaken by their R&D subsidiaries. The flow of innovation is clearly defined as 'from west to east'. Only if this paradigm can shift and the ideas of the subsidiaries are taken seriously there is a chance for reverse innovation.

To change this direction requires a major shift in thinking: From a colonialistic, ethnocentric view to an open, geocentric perspective. "Designed in China is not our tradition!", if firms retain such a mindset it is only a matter of time before innovative Chinese competitors take over their markets.

Rethinking Business Models Established firms often have old, and in many cases, proven business models. Their businesses have worked that way for decades. The management has a tough time proclaiming that the firm will market lower-price products developed in emerging countries in the future. It is a challenge to overcome the resistance of the R&D-departments in the western countries and to include them into the process.

Business models have to be adapted in terms of market segmentation, too. Low cost operating rooms for surgeries serve a need of Chinese hospitals. European hospitals have different standards and form as such a completely different market.

However, mobile operating rooms, such as Switzerland's 'mobile hospitals', might open up new opportunities for Chinese manufacturers. Rethinking business models and adapting customer needs becomes a key success factor for Chinese exporters.

Incentives for Local Product Development Both, western R&D-labs and the R&D-labs in emerging economies must have incentives to develop. It is a successful practice to market both kinds of products. Such a strategy fosters internal competition and thus ensures that new products are being developed.

Often the absence of sanctions and guidelines for using corporate components is enough to foster innovation in local subsidiaries. Once a product has been developed successfully, one challenge remains: How to protect the intellectual property that has been created in China?

Protection of Intellectual Property in China

2

Yet, there still is a dark side to innovating in China: the protection of intellectual property. The phenomenon of counterfeits originating from China has increased constantly over the past two decades. Moreover, within the past 10 years the scale of intellectual property theft has risen exponentially in terms of its sophistication, its volume, the range of goods, and the countries affected (ICC 2006). China's output of imitations was almost developed in parallel with legitimate manufacturing, producing and distributing an estimated 65–70 % of globally all fake goods (ICC 2007).

In 2009 EU customs seized more than 118 million counterfeited and pirated products leading to the assumption that the production of fakes in large quantities is likely to continue (EUC 2009). The range of goods has also extended to various industries. For example, although textiles remain the most intercepted counterfeit product, the manufacturing and industrial goods industries constitute an increasing threat with regards to pirated products. According to a 2007 survey of the German Engineering Federation (VDMA), 67 % of the respondents (manufacturing and industrial goods firms) claimed to have suffered from product piracy. In 60 % of the cases entire machines were the targets of imitation, while in 42 % of the cases spare parts had been copied.

Globally, three quarters of all imitations originate from China. But counterfeits in China go even further, only recently an entire group of 'fake Apple stores' was discovered. These stores did not only sell fake goods, they imitated the entire look and feel of the original stores. Store layouts matched those of the original stores, so did the employee outfits and the promotional posters on the walls. The same happened to IKEA. In Kunming in the southwest of China an IKEA store can be found—only it is not IKEA. The counterfeiter copied every aspect of an IKEA store, the layout, the small pencils, the products and even the yellow shopping bags. The store is called 'Shi Yi Jiaju' IKEA's Name in China is 'Yi Jia-Jiaju', one has to look twice to see the difference.

In this context, the protection of intellectual property and their associated rights remains a challenging task for any industrial firm. The development and elaboration of effective and solid protection strategies, including legal and factual protection means, need to receive careful managerial consideration.

With its entry into the WTO in 2001, China triggered the enforcement of several intellectual property laws to comply with the obligations of the Agreement on Trade Related Aspects of Intellectual Property Rights (TRIPS). However, firms still face risks concerning intellectual property when developing and manufacturing in China due to the lack of transparency within the legal system and the weak enforcement of intellectual property rights.

China currently ranks among the fastest growing economies in the world. According to a Goldman Sachs study (2004), China's Gross National Product (GNP) will exceed that of the United States by 2040. Harvard professor and well-known historian Niall Ferguson goes even further. In his current book 'Civilization—The West and the Rest' (Ferguson 2011). Ferguson claims that China's Gross Domestic Product (GDP) will surpass that of the U.S. within the current decade. Everywhere on the planet, companies embrace this opportunity by entering the Chinese market, either through wholly owned subsidiaries or by joint ventures. Concurrently, the Chinese government attempts to build its economic cornerstones not only on its low-cost manufacturing capability, but also on technology- and knowledge-focused industries. In 2011 these efforts to promote innovation and protect intellectual property have been intensified by the Chinese government.

In a survey conducted in 2004, the Delegate Office of German Economy interviewed 243 companies regarding their reasons to invest in China. Among the companies consulted, 94 % named 'future market access' to be the main motivation behind their investments. An additional 46 and 42 % indicated low production costs and the necessity of following major customers as motives for investing in China respectively.

Depending on the branch, the supplier relies on its clients' supporting market access. In the textile industry, for example, all major players are located in China or other emerging markets—very few international companies are still to be found in Europe. The electronics industry—as well as the biotechnology and pharmaceutical industries—anticipates similar developments. With an enormous capacity of low cost workers, China has literally become the world's shop floor.

Due to the attraction of direct foreign investments, the strategy of exchanging technology for market share has been established and supported by the Chinese government during the past two decades. Innovations in China are no longer exceptional cases but increase as a result of China's R&D spending. Today, China is already the worldwide leader in number of people engaged in science and technology. In 2009 the country accounted for 6 % of all scientific articles published (up from 1.6 % in 1995) and thus ranks at the number five spot globally (EIU 2009). University graduates with degrees in science and engineering represent 40 % of all Chinese graduates. That is almost twice the OECD average and it drastically exceeds the 15 % recorded in the U.S. (EIU 2009). The innovation index of the Economist Intelligence Unit (2009) predicts that China's innovation performance has and will outpace other emerging economies, improving by 11 % and jumping from 54th to 46th place comparing 2004–2008 and forecasted data 2009–2013. In comparison to other BRIC countries, India will rise up four places; Brazil and Russia remain unchanged (EIU 2009).

With the building of modern production and distribution, the imitation of goods has become far easier. Subsequently, the market for imitations has developed an organized criminal dimension.

The success formula for product and technology pirates is simple and follows a rationale:

Economic success of violators	= low costs
	+ high margins
	+ large markets
	+ low quality standards
	+ soft IP defense

As long as the domestic demand for intellectual property protection remains relatively low, the effective legal protection of a firm's intellectual property will most likely remain weak (Fuchs 2006). However, as long as market access outweighs the potential risks, investments in Chinese production and R&D sites will continue to increase in the future.

International firms are setting up production as well as R&D sites in China, such as the German consortium of Thyssen Krupp and Siemens. They developed the first commercial magnetically levitated train (maglev) for a line between the Shanghai International Airport and the city's financial district. In 2003 the project of the Transrapid train was established by Thyssen Krupp and Siemens conjointly with a Chinese consortium, whose leadership was connected with the Shanghai city government. Three years later, the Chinese Aviation Industry Corporation (CAC) introduced a maglev prototype on a track at the Tongji University in Shanghai. The short development cycle raises the question of intellectual property and know-how transfer or leakage. In that case the fundamental patents had already expired; however, the drainage of know-how remained. Thus, the effectiveness of patents, trademarks, utility, and design models may offer a protection of R&D efforts but depend on their duration and on their enforcement in case of an infringement.

Another example of increasing R&D efforts in China is ABB. In 2007 ABB announced that it would be delivering automation and electronic systems to 74 vessel cranes in China, an order totaling USD 65 million. The vessel cranes will be built by the Shanghai Zhenhua Port Machinery Company Ltd. (ZPMC). ABB is responsible for the entire project management, engineering, costumer training and commissioning. During the past decade, ABB has been considerably engaged in the Asian region, seeing a growth in its Asia sales from 14 to 23 % over the period from 2002 to 2005. In 2005 ABB established two R&D centers, one in Beijing and the other—the new ABB Robotics headquarters—in Shanghai. Local human resources

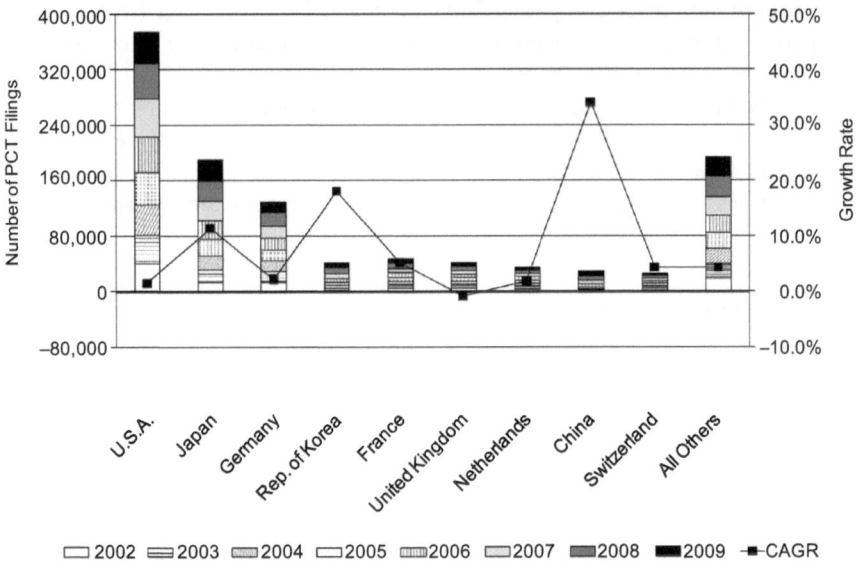

Fig. 2.1 PCT applications 2002–2009 with the highest compound annual growth rate of applications by Chinese firms

and the immense market have been the main drivers for investments by ABB. With 366 patents granted in China in 2006, ABB follows an active intellectual property rights management in China. However, due to the weak enforcement of rights, the firm's key strategy of protection relies on their service, business and innovation capacities. ABB's low tension switch that lacks legal protection by ABB is a common target for infringers. ABB actively forestalls and combats imitations supported by its production, distribution and sales team.

But, the Chinese patenting behavior is changing. Today, China ranks after the U.S., Japan, Germany and Korea on the 5th position globally. The Chinese tele-communication equipment firm Huawei, was the top patent filer in 2008 (and the number two in 2009 trailing Panasonic). While most countries (and the world on average) registered decreases in patent filing in 2009, China's rate of patent growth remained around 30 % last year. China is filing ten times more patents than its emerging market competitor India.

The application of IPRs is vital to safeguard own developments. The relevance of patents as the strongest intellectual property protection right has been recognized by Chinese firms represented in the highest CAGR increase of 34 % of PCT applications (Fig. 2.1).

Case Example IKEA

Chinese companies not only plagiarize unique products. IKEA is facing the threat of entirely fake stores. The Chinese furniture store '11 Furniture' (which in Chinese sounds roughly like IKEA) looks and feels almost exactly like the Swedish brand. The store is designed around a maze of furniture. Small wooden pencils are provided to help you write down what you want to pick up later. Blue shopping bags with yellow handles (see picture) can be used for all smaller items.

© Reuters

Once you made it through the maze of furniture you enter a second maze. Kitchen utensils, boxes, carpets, decoration objects and much more are presented in the exact same way as in original IKEA stores. Even the price tags and the shopping carts are the same (see picture below).

© Reuters

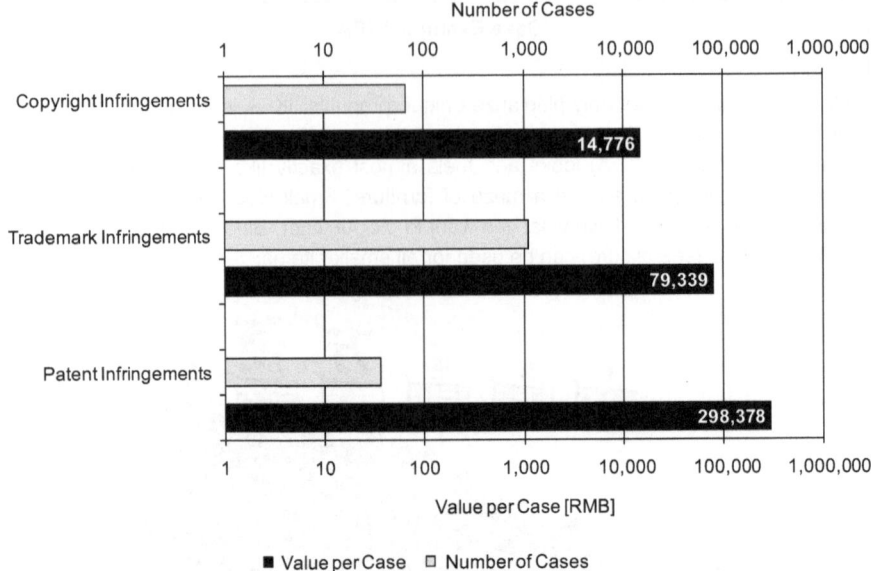

Fig. 2.2 Infringement cases registered by Chinese customs. (Source: Data from GAD (2005))

Chinese technology leaders such as Lenovo, ZTE, Haier and Huawei increasingly apply patents in China and abroad. They have a vital interest in an enhanced protection and enforcement of intellectual property rights. According to the Patent Protection Association of China (PPAC), the three Chinese firms ZTE, Huawei and Hongfujin have applied the highest amount of patents in 2009. The telecommunication firm ZTE increased their patent applications by 20 % compared to 2008 and ranks first with 5,719 patent applications. The firms' foreign patent applications increased by 200 % or respectively 1,164 patent applications. In contrast to the high amount of Chinese utility model applications applied by local firms, more than 90 % of ZTE's applications account for patents. ZTE's strategy for the next years is to increasingly invest in R&D and its protection by means of keeping the level of 10 % of revenues investments for R&D and its protection.

Despite the growth of intellectual property rights in China, the imitation of goods remains a challenge for domestic and foreign firms alike. The complexity in terms of the professionalization of illicit production of goods has increased considerably. According to Chinese Customs, registered infringement cases have increased from 193 to 1,210 between 1997 and 2005. Trademark infringement has increased from 557 to 1,106 cases between 2002 and 2005 (Chinese Customs 2007). Similarly, copyright infringements increased from 2 to 67 cases and patent infringement increased from 14 to 37 cases during that same timeframe. Although trademark infringement constitutes the largest amount of cases, patent infringements have accounted for the highest value per case (see Fig. 2.2).

A study of the German Engineering Federation (VDMA) notes that Germany has become home to the second largest machinery and industrial goods industry worldwide with exports reaching a record high of EUR 123 million in 2006. Concurrently, product piracy has been ranked as the largest threat to the industry.

Seventy six percent of the respondents (manufacturing and industrial goods firms) claimed to suffer from product piracy. The need for a fresh approach is necessary for European firms to take advantage of China's explosive domestic market while not exposing its intellectual property to continued losses. Legal and factual protection means may enable a better protection, but the question remains how to derive an effective protection strategy to preserving one's intellectual property while prospering in the world's largest marketplace.

2.1   Enforcement of IP-Rights in China

IPR litigation is generally lengthy, complicated and costly. In China the patent system is relatively new. Litigation and enforcement is based on the Civil Procedural Law that governs the procedures for patent litigation, the Patent Law and the Implementing Regulations of the Patent Law. The legal system does not have any case law but statutes, which are binding. Judges make their decisions on a case-by-case basis according to their interpretations of the law.

This system can result in unpredictable trials, especially within courts with little experience. Moreover, the only recognized binding precedents are the so-called judicial interpretations from the Supreme People's Court, the highest of the four courts (Cheong 2006). These decisions provide guidance to lower courts and the Supreme People's Court follows its previous decisions. Furthermore, the Lower Courts cannot deliver judgments that are obviously contrary to a previous judgment of the Supreme People's Court. Thus far, several judicial interpretations on IP laws have been issued; in addition, some responses to requests for clarification by lower courts have been established by the Supreme People's Court for important guidance.

The **judicial system in China** has four tiers of courts:

1. The Supreme People's Court
2. The High People's Court
3. The Intermediate People's Court
4. The Basic People's Court

The Supreme People Court is the highest judicial authority of the state. It is located in Beijing. Its interpretations are legally binding although there is no formal rule of stare deceases. A High People's Court is located in every province, autonomous region and municipality directly under the central government. Furthermore, there is at least one Intermediate People's Court in each medium-sized city directly under the central government as well as one Basic People's Court in each county.

Concerning IP matters, there is no specialized IP Court but specialized divisions for IP cases. The first specialized division was established in Beijing in 1993. Three years later, the Supreme People's Court set up an IP division, and by 2001, every high court and many intermediate courts in major cities followed suit. Generally, the court of first instance for a patent infringement is the Intermediate People's Court. By the end of 2007 64 Intermediate People's Courts had exclusive jurisdiction over patent cases. Due to the established two-instance procedure, the decisions of the court of first instance can be appealed to the next higher court, which has the final decision. In addition, if material is proven incorrect by evidence, the higher-level court or even the Supreme People's Court can retry the final case, where both issues of fact and law will be reviewed (Jian 2007).

The enforcement of infringed legitimate rights can be enforced by means of a so-called 'administrative track' or 'judicial track'. The judicial track includes the civil and the criminal, although the latter is of negligible meaning in most cases. The choice of track depends on the various types of cases. Both tracks can be initiated yet not simultaneously. Thus, a concluded administrative track can be followed by the civil track.

Globalization, in particular the emergence of China as a formidable competitor on the global stage, means that companies across the world increasingly find themselves entangled in complex 'webs' of relationships (Pisano 2006). In the context of China, appropriability regimes are weak and innovation draws rapid imitation (Pisano 2006). A profiting from innovation (PFI) framework helps organizations to systematically deliberate the kinds of assets they need to foster internally and those that they can safely outsource (Pisano 2006). Nature of knowledge, intellectual property protection, and asset structure of a firm impact upon the business enterprise's ability to capture value from innovation. Appropriability mechanisms reflect the embedded ability to return profits from innovations.

2.2 Appropriability Regimes and Their Mechanisms

Appropriation of returns from technical innovations is very important for individual inventors and innovators as well as for technical change in individual markets and the economy as a whole (Harabi 1995). The appropriability regime has become a recognized concept in strategy (Teece 2006). In 1986 Teece's framework for profiting from innovation—the most cited article in the leading journal 'Research Policy' of all times—introduced the hypothesis that appropriability, and success at innovation in general, is related not so much to the innovator's ex ante market share but to the complementary asset structure of the innovator, the management's market-entry timing decisions and the contractual structures employed to access missing complementary assets. Choices with respect to the latter should depend on the asset positioning of other market participants and on the intellectual property protection available. Profits from innovation depend upon the interaction of three preeminent factors, namely appropriability regimes, complementary assets and the presence or absence of a dominant paradigm. Importantly, Teece (2006) represents intellectual property protection as just one among many barriers to imitation. He also refers to the nature of knowledge—especially the degree to which it is tacit—and its inherent duplicability as another imitation barrier.

Harabi (1995) reveals the difference between appropriability ex ante and appropriability ex post. The ex ante notion emphasizes the potential capability of an innovator or the organization which owns the innovation to fully, or at least partially, appropriate the returns from innovation. Having a critical mass of appropriability capabilities acts as a market-entry barrier and, therefore, as a protection from competitors.

Harabi (1995) groups the major **means of appropriability** into three sub-categories:

1. Patents
2. Secrecy
3. Lead-time and related advantages
   (so-called first-mover advantages)

Lead-time could be used to strive for further advantages in manufacturing by achieving and securing learning curve advantages and in marketing by means of building up superior sales and service efforts. Lead-time could also be used to either hinder or at least delay the imitation by competitors of a firm's own innovation (for example by increasing cost and imitation time). Conversely, patents are considered on average as ineffective in securing licensing fees (Harabi 1995). Furthermore, the effectiveness of patents in preventing duplications was identified as being low (Harabi 1995) for both process innovations and product innovations.

Limitations on the effectiveness of patents for protection are well known:

1. Not all new and improved products are readily patentable
2. If challenged, patents can lose their validity
3. Firms do not attempt to enforce patents
4. Competitors can legally invent around patents
5. Technology is moving so rapidly that patents become irrelevant
6. Patent documents require disclosure of too much information
7. Legal limits on licensing
   (mandatory registration, compulsory licensing, etc.)
8. Cooperation in R&D projects, also between competitors
   (see Levin et al. 1987)

Firstly, Harabi's (1995) studies indicate that the ability of competitors to invent around patents is the most important limit to the effectiveness of patents as a means of protecting competitive advantage of innovations. Secondly, legal restrictions on licensing and cooperation with competitors are the least important constraints on the effectiveness of patents.

Yet, further means of patents include:

1. Disrupting R&D or the product lines of competitors
2. Evaluating the performance of R&D employees
3. Achieving/maintaining desirable negotiating positions with other firms
4. Entering foreign marketsa
   (either through direct investment and production or indirectly through granting a licensing agreement)

Harabi (1995) identified secrecy as a relatively effective means of protecting process innovations against imitation with the exception of the construction and watchmaking industries. On average, however, experts within the Harabi survey rated secrecy as a relatively weak means of appropriability.

Furthermore, capturing and protection cost advantages as well as superior sales and service efforts were identified as relevant to process innovations but only if these are relevant for product innovations and can be marketed as such. Generally, this method of appropriability was assessed in all industries as being very effective.

The importance of inter-industry differences was reflected in the effectiveness of the different means of appropriability. Complex product industries, in which a product is protected by a great number of patents, generate and use patents to rival use of components and acquire bargaining strength in cross-licensing negotiations. In contrast, discrete product industries, in which a product is relatively simple and therefore associated with a small number of patents, use patents primarily to block substitutes by creating patent fences (Gallini 2002; Ziedonis 2004).

Harabi (1995) recommends that firms, which in market economies are the main actors in the innovation process, implement a strategy capable of protecting their innovations. They are advised to design a patent policy that takes into account the technical nature and life cycle of their products, as well as existing market conditions and structures (Teece 1986). Mansfield et al. (1986) expound first that patents entail significant imitation costs. A firm's intellectual property portfolio cannot be managed independently of its business strategy and that business strategy formulation requires an appreciation of intellectual property issues (Teece 2006). The 'Profiting from Innovation' framework is a robust framework for helping to explain why firms and individual entrepreneurs are supposed supporters of strong intellectual property while certain large firms may be indifferent or possibly even opposed to intellectual property protection. Larger firms are likely to have established processes to protect their own intellectual property. Alternate strategies for capturing value from their own and other's innovations are more likely to be embedded.

Although this framework has initiated the concept of appropriability regimes and appropriability mechanisms, the challenge of strategy is to develop appropriate vertical integration and complementary asset positions given the existing appropriability regime (Pisano 2006). The understanding and evolution of appropriability regimes has not been the subject of deep strategic management yet and poses the question about how much firms can proactively shape an appropriability regime in their favor (either by weakening or strengthening it) and what the mechanisms are that firms can use to shape an appropriability regime (Pisano 2006). The proposed thesis endeavors to elaborate on mechanisms that firms can use to strengthen their protection and appropriation.

Innovators and imitators share similar search processes—doing, hiring, reverse engineering and other forms of search (McKendrick 1995). Imitators are not bound by the same constraints of innovation. For imitators, the solutions to analogous problems or opportunities will have already been discovered elsewhere and the successful routines of innovators can be observed, albeit imperfectly. For imitators the target is more visible and the direction of search more certain. Imitators draw on a variety of already-developed external sources such as competencies embedded in consultants or other firms and information in the public domain. Thus, imitators seem to have

Case Example: Accessories

The main threat of Chinese counterfeits often comes not in the form of the actual product but in the form of its accessories. Chinese companies sell accessories that resemble the original ones to a degree that makes it challenging for customers to tell them apart. Phone manufactures see this as a major challenge. The fake accessories are mostly labeled as "originals". In many cases their poor quality casts a shadow on the original manufacturer as customers complain about quality issues. The following picture shows two almost identical Nokia phone chargers. Both chargers are rated 5.0 V/890 mA. However, the counterfeited charger has an actual output that fluctuates between 5.0 and 9.5 V. It basically fries the cellphone circuitry. Yet, the customer did not even know that he was buying a counterfeit. The more standards are designed, the easier and more attractive it is for imitators to break into the business. The standardization of chargers might be the right thing for the customers, but it is also a great opportunity for counterfeits.

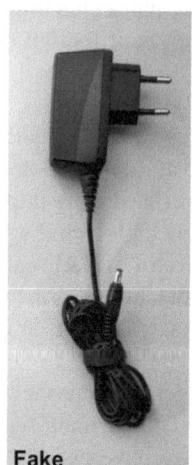

Original   Fake

Even more daunting is the situation in the machine building industry. Chinese companies offer after-sale services or spare parts at far lower rates than the original products. However, these spare parts often do not meet the needed quality standards. In many cases the use of these spare parts has a critical effect on the entire machine. In one case for instance a Chinese firm (under the label of SIEMENS) sold gear wheels for gearboxes. The gear wheels were made of low quality steel and not carburized. After only a couple of weeks they broke and wrecked the entire gearboxes. The customer was stunned when he found out that he had bought counterfeits. The damage was many times as high as the price for original accessories.

the potential to tap a greater multiplicity of sources in their search to replicate or approximate know-how. Therefore, the increase of imitation barriers is essential for IP protection.

Mansfield (1981) suggested raising the imitation costs and time for a stronger appropriation of innovations. McKendrick's study (1995) discovered that professionals in the banking industry carry their knowledge from company to company, which improved the managerial and technical capabilities of Indonesian firms. Informal sources such as training or hiring staff from other firms are identified as effective means for accumulating process know-how.

2.3 Patents as an Indicator of R&D Capabilities

A singularity about China's IP system is the 'Indigenous Innovation Program', which entitles Chinese (but not foreign) firms to receive tax credits and premiums for each patent they register at the SIPO (Zhang et al. 2009). Many firms emphasize that this policy has a side effect: patents are seen as an indicator of the R&D capability of a firm, be it foreign or domestic. Thus, two effects for foreign firms can result:

- *Firstly*, firms report that signaling the status as a high-tech company by filing patents helps to attract domestic business partners when collaboration is needed.
- *Secondly*, and perhaps more importantly, not having patents indicates that a firm is not involved in high-tech. Firms explained to us that patents are seen far more often as an indicator of R&D capabilities as that is the case in western countries. It is unlikely that Chinese firms will consider a western firm as R&D intensive if that very firm has no patents in China.

Some R&D managers we interviewed even explained to us that this signaling quality of patents still is their main reason to patent in China at all. High-tech simply cannot afford not to have patents in China, the R&D managers pointed out. The Chinese Patent Office currently encourages PCT patents (more than five countries) with 65,000 €. In the following paragraphs we will explain the tax credits and premiums Chinese firms can apply for in order to provide western R&D managers with an overview.

Scope of **funding**:

1. Application Fee, substantive examination charge and service charge for domestic patent, utility and design model filing,
2. Other charges that the fund management body deems to be funded,
3. Partial filing charge of patent filing in foreign countries via PCT or other channels,
4. Any service charges for patent filing, which are included in the Government projects, shall not apply for the fund again.

Scope of **Standard**:

1. Patent filing in foreign countries via PCT, funding can be made on different phases respectively: RMB 10,000 per case for international phase and RMB 10,000 per case for the national phase.
2. Patent filing in foreign countries via other channels will be funded with RMB 20,000 per case and country.
3. When filing in multi-countries, fund can be made for filing in max. five countries. Each unit can obtain max. RMB 500,000 as filing fund.
4. From Jan. 1st to Dec. 31st of a year, for unit applicants, who filed more than ten patents in foreign countries (including ten cases), extra RMB 10,000 will be funded to each granted patent.

Dimensions of IP Protection in China

3

The application of IP protection means depends on the type and the characteristics of an IP threat. Common IP threats in China concern legal infringements such as patent, trademark, design or copyright infringements. The OECD (2007) examined the legal infringements of trademarks, copyrights, patents and design rights to the extent that they involve physical products. Recent research also refers to the WTO (2004) definition of counterfeit trademark goods and pirated copyright goods (OECD 2007; Staake and Fleisch 2007)[1]. These legal infringements have been identified on the demand side as well as on the supply side (Bloch et al. 1993; Bush et al. 1989; OECD 2007; Tom et al. 1998). IPR infringements are only part of the IP threat environment in China. In addition to such legal-illicit threats, also legal-licit, that is to say uncertainties within the boundaries of the law and factual challenges are predominant in China.

3.1   Legal IP-related Illicit and Licit Threats

The varieties of legal threats are manifold and usually vary in each particular case. *Legal-illicit* threats include infringements of IPRs, trade secrets, contractual agreements and relate to a technical product, process or related know-how. The legal-illicit threats are a direct infringement of an IPR or of a contractual agreement. *Legal-licit* threats are challenges that are within the boundaries of the law but still

---

[1] The Agreement on TRIPS provides a definition for counterfeiting and piracy, in a way that: (a) counterfeit trademark goods shall mean any goods, including packaging, bearing without authorization a trademark which is identical to the trademark validly registered in respect of such goods, or which cannot be distinguished in its essential aspects from such a trademark, and which thereby infringes the rights of the owner of the trademark in question under the law of the country of importation; and (b) pirated copyright goods shall mean any goods which are copies made without the consent of the right holder or person duly authorized by the right holder in the country of production and which are made directly or indirectly from an article where the making of that copy would have constituted an infringement of a copyright or a related right under the law of the country of importation.

O. Gassmann et al., *Profiting from Innovation in China*,
DOI 10.1007/978-3-642-30592-4_3, © Springer-Verlag Berlin Heidelberg 2012

concern IP and are difficult to counteract due to the absent legal enforcement. The protection of products and know-how that is not legally protected is very difficult. Imitations based on unprotected technologies, business models, and service concepts are particularly challenging.

Furthermore, the very number of Chinese IPRs is a challenge when doing business in China. Unknown IPRs represent a major threat since they bear the risk of infringing third parties' rights. Further legal-licit threats are related to legal regulations and usually complement illicit threats, such as the uncertainty of success in litigation. They are based on challenges and anticipated risks within China's IPR appropriability regime.

Factual Challenges Related to IP Protection A broad range of factual challenges in IP protection in China exists. These challenges concern IP protection in China but have no legal foundation. Examples of factual challenges are mainly based on differences in culture, loyalty, and commitment to the firm, an understanding for the importance of IP or privileges shown to local competitors and local protectionism.

In the following section the legal-illicit IP-related threats, mainly IPR-related, are described followed by the legal-licit threats and factual challenges.

Invention Patent Infringement The manufacturing and industrial goods industry is characterized by technological innovations that are often covered by patents and industrial processes, some of which are patent-protected. Patent infringements arise as a result of the exploitation of a patent without the authorization of the patentee. Since patents are public documents, they are constantly at risk of being copied or otherwise infringed.

Reverse engineering of technical products and parts are one means to examine how the new technology works or how a technical part was produced. Reverse engineering may also be used to identify whether a firm's own patents have been infringed by a competitor. Despite this, the case studies show that reverse engineering is a limited tool in this industry since the products are difficult to obtain given their subject matter.

Trademark Infringements The available evidence is that the copying of trademarks constitutes the greatest proportion of IPR infringements in the manufacturing and industrial goods industry. Trademark infringements involve the use of a trademark that is identical or similar to a registered trademark without the authorization from the trademark registrant. It also includes the selling of goods that knowingly bear a counterfeited registered trademark or the making or selling of a representation of a registered trademark. The replacement of a trademark and the subsequent marketing of the good, which carries the replaced trademark of the original trademark registrant is considered a trademark infringement (if the original trademark registrant has not consent to it). In the machinery and industrial goods industry the most common IPR violations are trademark infringements. Trademarks are used to illicitly exploit a well-recognized reputation of quality and reliability associated with right holders' products.

Design Patent Infringements The infringement of a design patent emerges once the patented design—as shown in the drawings or photographs of the applications—is exploited, that is made, sold or imported into a product that incorporates the design without the authorization of the right holder. Therefore, in a design patent infringement, the objects of manufacturing, selling or importing are products that incorporate a design, which is patent-protected. The product that incorporates the design must be identical or similar to the patented product and the design must be identical or similar to the patented design.

In the manufacturing and industrial goods industry, design patent infringements have been perceived least relevant. This can be explained by the subject matter, which can be legally protected. In this industry the protected technologies differentiate from competitive products by means of technical and functional features and their applications rather than their design. If the design has a significant impact (for instance by means of the reputation) or has a positive impact on the customer value (such as the handling of a product), the unauthorized exploitation of a design might be of higher relevance and the correct design protection should be considered.

Copyright Infringements Copyright protection covers the right of publication, authorship, alteration, integrity, reproduction, distribution, rental, exhibition, performance, showing, broadcasting, communication, making cinematographic work, adaptation, translation, and compilation of works. The term "works" includes works of literature, art, natural science, social science, and engineering technology.[2]

Trade Secret Infringements A trade secret is a non-registrable intellectual property, which is technical and/or management information that is unknown to the public, that brings an economic benefit to the owner and for which the owner has adopted measures to maintain its confidentiality. The protection against the fraudulent disclosure of a trade secret is defined in Art. 10 of the Anti Unfair Competition Law of the PRC and comprises protection against following acts:

1. Obtaining the trade secrets of any rightful party by theft, inducement, duress, or other illegal means.
2. Disclosing, using or allowing others to use the trade secrets of any rightful party obtained by illegal means.
3. Disclosing, using or allowing others to use the trade secrets in breach of an agreement or the confidentiality requirements imposed by any rightful party.

---

[2] Copyright Law of the People's Republic of China, Chap. 1, Art. 3 specifies the term 'works' as literature, art, natural science, social science, engineering technology and the like which are expressed in the following forms: (1) written works; (2) oral works; (3) musical, dramatic, quyi', choreographic and acrobatic works; (4) works of fine art and architecture; (5) photographic works; (6) cinematographic works and works created by virtue of an analogous method of film production; (7) drawings of engineering designs, and product designs; maps, sketches and other graphic works and model works; (8) computer software and (9) other works as provided for in laws and administrative regulations.

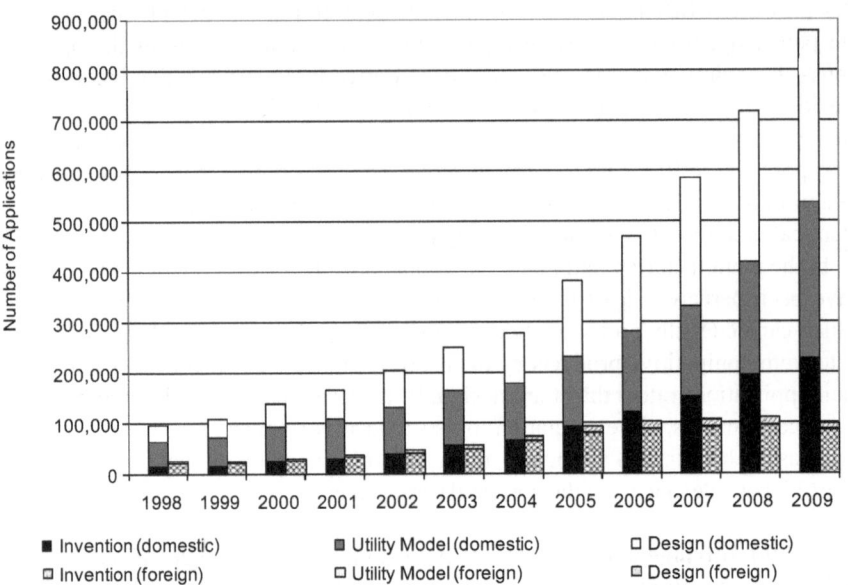

Fig. 3.1 Domestic and foreign applications of three kinds of patents in China (1998–2009). (Source: SIPO (2010) data)

Also, third parties who obtain, use or disclose trade secrets that they know are infringing acts according to 1.–3., will be deemed to have infringed the trade secrets. The manifold varieties of trade secret infringements in China are often related to know-how theft by employees, former employees or employees of suppliers and partners.

Own Risk of Unknowingly Infringing Third Parties' IPRs The risk of infringing third party patents always exist if competitors apply patents for technologies and innovations in countries that a company operates in. In China this risk has been identified as more severe than in other countries due to the strong increase in the amount of patent and utility model applications and grants in existence. With a total amount of 337,215 patents in force (WIPO 2010), China is the third largest country in terms of patents in force worldwide (Fig. 3.1).

The amount of utility model applications is almost as high as the number of invention patent applications, rising to 310,771 in 2009, with domestic applicants representing a 99 % share of this total. Utility models only exist in Chinese and create a vast landscape of potential IPRs that could be violated by third parties. Due to current developments and decisions, the infringement of third party patents in China could present an expensive and risky obstacle. In particular, the trend for offensive patent enforcement creates a worldwide potential for the offensive use of own IPRs against third parties. The Chinese electronic manufacturer Chint has enforced its utility model patents against the Chinese subsidiary of the French Schneider Electric company. Chint is one of Schneider's major Chinese competitors. Schneider has battled with Chint on several continents to restrain Chint's presence in the global marketplace (Foley 2009). The French company has initiated several infringement cases against Chint in Europe and globally over the past 20 years. In 2006 Chint accused Schneider Electric Low-Voltage (Tianjin) Co., Ltd. (Schneider) of infringing one of its utility model patents, demanding RMB 500,000 Yuan in compensation. In the same year, Chint initiated a lawsuit in China with the Wenzhou Intermediate People's Court against Schneider. During the 2 years of legal proceedings, Chint defended the validity of the concerned utility model patent before the PRC Patent Reexamination Board and succeeded in winning the patent infringement lawsuit at the Wenzhou Intermediate People's Court. The amount of RMB 334 million Yuan represented the largest amount ever rewarded in China for a patent infringement. Subsequently, Schneider appealed against the lower court's ruling in the High People's Court in Zhejiang. In April 2009 Schneider settled its patent lawsuit with the Wenzhou-based Chint for RMB 157 million Yuan (approx. USD 23 million).

The offensive enforcement of patents has increased within the past decade with companies aggressively enforcing acquired patents for the generation of licensing revenues. The so-called 'patent trolls' have begun to enforce patents often without any own intention to manufacture or market the patented invention. This approach is legally conform and relates to an interfirm enforcement of rights; but since patent trolls do not intend to manufacture, cross-licensing solutions or negotiation powers over other IPRs are not available. The creation of patent walls or patent thickets is another means created to defend against competitors designing around a single patent (Shapiro 2001). In the Chinese context, the utility model patent thickets carry the risk of infringing third parties' patents. This presents a challenge when operating in China since the costs for the scanning, monitoring, identification, and translation of utility models often exceed the IP budget.

Contractual Breaches by Suppliers and Partners The contractual breaches by suppliers and partners in terms of IP-related issues occur through the misuse of confidential information and breaches of terms which had been stipulated by contract. Both forms are difficult to identify and to prove. Common contractual breaches include the sales of confidential information by the employees of cooperation partners. Over-production or so-called factory overruns are also considered contractual breaches. One can presume that the main incentives for such breaches are based on monetary or political grounds. The networks of illicit actors in such cases are difficult to identify. The collection and proof of evidence of a contractual breach are the main challenges of such illicit incidents.

Obscurity of Criminal Networks and Logistic Chains The underlying networks of imitators are often based on organized crime and obscure political as well as social networks. Although the business centers of the Chinese economy are located on China's eastern coastline, the country's vast size facilitates the existence of an obscure landscape, which bears unknown factories of imitators.

Previous studies have analyzed product piracy and the logistics chain and have seen that the identification of the imitators may not allow the enforcement of proprietary rights, particularly if evidence cannot be linked with the infringement and the underlying networks.

Parallel Trade Parallel trade concerns imports of non-counterfeit goods, which have been distributed by the rights holder in a different country and are imported without the permission of the IP rights holder to another country. Parallel imports occur as a result of disparate price strategies in different countries and for different versions of a product for the various markets. Parallel imports are difficult to detect at the border since the goods are no different to the legally imported goods. The principle of territoriality of IPRs may limit the legal protection means in the import country if IPRs have not been applied or once the goods are distributed in the market. Due to the direct distribution channels or distribution by retailers in the manufacturing and industrial goods sectors, parallel trade is less relevant due to the subject matter of industrial goods which have to be manufactured in China or due to the pricing and product portfolio strategies of the companies interviewed.

3.2   Legal-Licit Related Threats

Circumvention of Protected Technologies and Products The circumvention of protected products is one of the challenges that companies face in China. Often, the rapid imitation development process considers the barriers of IPR restrictions and the applied technologies, trademarks or designs are just within the legal boundaries. In China such licit imitation approaches are manifold and particularly challenging due to their regular occurrence and their conspicuousness.

Vast Amount of Invention and Utility Model Patents Available Only in Chinese Language The number of utility models and design model applications by far exceed the number of invention patent applications in China. Since these documents mainly exist only in Chinese, the monitoring of such large numbers of IPR documentation is time and cost consuming and often exceeds the IP budget. While invention patents are to some extent available in English, the utility models only exist in the Chinese language. Thus, firms are at risk of unknowingly or unintentionally infringing third parties' IPRs (see paragraphs on legal-illicit threats).

Another challenge has been perceived in IPR applications that might include the existing state of the art and have been issued inadvertently in China. Since utility model applications were previously not subject to profound examination reports, the existing state of the art might remain unconsidered. In contrast, invention patent applications are examined as to their substance. With the new Chinese patent law that came into effect in October 2009, the latter uncertainty should be reduced.

Former Definition of Novelty in China The requirements of the novelty definition and the related state of the art definition vary in different national IPR legislations. In European countries the definition of absolute novelty requires that inventions are not state of the art. The state of the art is defined as the existing knowledge prior to the priority date of the patent application, which exists in written or oral explanation, by means of usage or in any other kind of publication. In China, the absolute novelty as defined in western countries has been introduced only recently with the new amendments in effect since October 2009. Prior to this, the definition of novelty differed from the European understanding in terms of the prior usage. The worldwide prior usage of an invention was restricted to a prior usage in China or a worldwide publication.

The difference in the absolute novelty term therefore gave rise to the risk that any applicants in China could apply patents or utility models for invention, which had been publicly demonstrated by an inventor outside of China. Subsequently, the inventor could have been challenged or asked for royalties with respect to the Chinese IPR if the invention had been introduced to the Chinese market. The 2009 harmonization of the law raised the bar of novelty requirements of patent applications in China and now relates to that of common practice outside China. Patents applied under the former legislation remain a risk for unknowingly infringing third parties' patents.

Obscurity of Interpretations of Legal Changes The legislation in China needs to be more transparent and more channels need to be made available for companies, industrial associations and the public to understand, apply and to participate in legislation (China's Action Plan on IPR Protection 2008). In order to deal with problems in the IP legislation promptly and effectively, the revisions and judicial interpretations of IP laws need to be improved. The judicial interpretations are established guidelines based on former court rulings to support the interpretation of the IP laws.

These judicial interpretations mainly exist in Chinese. To make use of the interpretations as a foreign firm, IP law firms need to translate them for their application. Thus far, a transparent and public version has not been launched by the SIPO. Due to language barriers, the detection and recognition of changing regulations are difficult to follow up for patent attorneys located at European headquarters. Chinese IP law firms are required to support the understanding and usage of such interpretations.

Uncertainty of Success in Litigation Before entering a litigation procedure the plaintiff and defendant typically endeavor to solve the dispute out of court. The IPR laws also demand the communication and the problem-solving attempts out of court. Although the uncertainty of success in a promising litigation case is not a China-specific concern, the perception of the IP enforcement regime is weak. The weak enforcement regime thus constitutes the uncertainty of success in promising litigation cases.

Costs vs. Restricted Compensations for Litigation Success in China In the event of an IP infringement that can be legally countered, awareness of costs is essential for decision making. Entering litigation is very expensive mainly due to attorney costs. While litigation costs are high, the compensation fees are more likely to be limited to a certain amount. This fact was mentioned by several firms as being a reason not to enter litigation even if evidence was sufficient. In the event of enforcement by the administrative procedure, the fines are limited to three times the illegal earnings, or if there are no verifiable illegal earnings, no more than RMB 1,000,000 Yuan. Thus, the investment in litigation may not be worthwhile regarding the monetary sense. However, motives for litigation are often based on the deterrence of further infringements rather than of compensation damages.

Evidence for Infringements A main challenge is the collection of evidence in China. The legal basis is clearly defined: According to the patent law[3] the rights holder has to authorize the exploitation of a patent, that is make, use, offer to sell, sell or import the patented product, or use the patented process, and use, offer to sell, sell or import the product directly obtained by the patented process, for production or business purposes.[4] In theory, infringements are legally easy to identify since the exploitation is always linked to a technology or good. In reality, the identification of infringements is challenging: The patented inventions are often part of machinery or part of a process, which is not accessible and not visible. Due to objectivity reasons the People's Courts rather than the involved parties or their representatives are responsible for the collection and analysis of the evidence. In practice however, it is the plaintiffs who provide most of the evidence in the administrative procedures as well as the civil procedures for IPR enforcement.

For a successful procedure, the amount and quality of the evidence can be essential for the plaintiff. It should particularly establish a clear link between the infringing IPR and the sued party. The identification of the relevant parties can be difficult, since manufacturers and sales persons often work undercover and often change their names and locations. Evidence in such networks is thus usually difficult to obtain.

3.3   Patents and Trademarks Are Not Enough

Factual IP threats are challenges and risks that are related to non-protected technology and know-how. There is a variety of threats which is not IP-related but de facto concern the protection of IP in China.

Drainage of Know-How and Unprotected IP The theft of know-how has been anticipated as a major challenge for companies that establish production and R&D sites in China. Due to the intangible character of know-how, the drainage of non-documented know-how is difficult to detect and trace. The tacit know-how is pertained in the heads of employees or partners who exploit it deliberately or unknowingly. With the fluctuation of R&D employees to other firms, their exploitation of proprietary IP cannot be controlled.

Head-hunting activities targeting key personnel have been challenging for European firms. Many firms have an unwritten agreement not to headhunt each other's

---

[3] Patent Law of the PRC, Art. 11 defines the need for authorization by the right holder of a patent.

[4] For the authorization of the exploitation of trademarks, refer to Trademark Law of the PRC, Art. 52.

employees. Multi-national firms with IP departments in China have reported that IP experts are common targets for headhunters since their western experiences are fruitful for most Chinese firms to establish their own IP management. Moreover, the experience of foreign IP practices is vital for Chinese firms or other foreign firms situated in China. Since loyalty to the firm is lower than in western cultures, the commitment to stay within a company is limited. Some firms report that they experienced western employees to feel bad when they terminate their employment for another employer.

In contrast, a manager of a manufacturing firm in China reported that an employee left the firm and notified his project manager by email about his termination of employment. In the event that the person changes his address it may become extremely difficult to follow-up on the contract breach. In any case, the firm cannot assure that she/he will keep know-how confidential; plus this includes the challenge to find an adequate replacement in a short period of time. Cultural differences such as the importance of guanxi may confront an employee of either behaving loyal to the firm or loyal to his or her network. A German industrial firm reported that in a meeting with a state-owned company an employee was asked which side he is on and that he should bear in mind that he is a Chinese. Another employee had to investigate IP infringements and did not deliver any results. When the manager asked him, he behaved unlikely withdrawn. His project leader found out that he and his family were threatened due to the investigations he was asked to do. The manager changed the tasks of that employee with immediate effect.

Cultural Differences and Lack of IP Awareness The lack of IP awareness is the baseline for most inter-firm IP conflicts. Employees may serendipitously disclose confidential information to third parties without considering the consequences of such a disclosure. The cultural differences are deeply rooted. In Confucian values, the imitation of successful people is a very important aspect. Learning by imitation has had a long tradition within Chinese education. Siemens in China reported that the notion of self-responsibility is higher amongst western compared to Chinese graduates. Often, for local graduates the importance of self-responsibility and the own creation of IP and its protection is lower. In negotiations and cooperation with local suppliers, the negotiation style is different. Written agreements are less binding than verbal agreements and personal relationships. Moreover, differences in expectations often lead to misunderstandings and communication conflicts.

Western practices are known as being straightforward and demand to be adapted into Chinese negotiation styles to achieve own objectives or a conjoint result. In contrast, western managers recently started to question the loss of face and lack of IP awareness. They anticipated that Chinese partners might exploit the fact of common cultural differences such as losing face and apply it deliberately for an enhanced negotiation power. They abuse the weak IP awareness reputation for their benefits. Such behavior is difficult to detect in particular if negotiations take place without any locals from one's own firm, who are more familiar with local culture, language and nuances of behaviors.

Local Protectionism and Privileges Some European companies report the unfathomable privileging of local companies. The support of local firms by means of subventions for IPR applications may create disadvantages for foreign firms. The privilege of local firms of receiving credit grants has also been reported. The grant of credits for plant building, taxation and pricing strategies of suppliers often challenge western companies. Since China has clearly put forth to strengthen the local innovation activities within the next decade, these privileges may remain in the midterm and are difficult to counteract for western firms.

Demand of Disclosure of Know-How for Potential Project Acquisition To stay competitive the pressure in China to disclose information before entering a business deal is exploited by some state-owned firms that are trying to gather as much information upfront. Siemens China reported that they were requested to disclose their blueprints in order to stay competitive in a public tender. In particular if assurance to win such tenders is open, the disclosure needs careful evaluation. Instal several times experienced receiving orders for industrial installation three or four times from a state-owned company, which afterwards rebuild the industrial installations and used the acquired knowledge solely. They clearly demanded a technology and know-how transfer in the first place which was then exploited.

This strategy meets the Chinese governmental technology plan to build up new technologies in China and support it by technology transfer from abroad. To enforce this strategy, the international competition put pressure on the applying firms to disclose more information than their competitors to win the deal.

Copy of Unprotected Technologies or Process Know-How The exploitation of unprotected tangible assets can also be a result of reengineering efforts of competitors or parallel development and improvements of technologies and processes by several third parties. An unprotected status can exist due to state-of-the-art technologies and processes, which cannot be protected by IPRs or due to a secrecy strategy or a limited IP budget. State of the art is either based on common knowledge due to former publications or the public usage and documentation of a certain invention. It may also relate to an abandoned patent or utility model. The fundamental patents of the maglev train are expired already, thus the exploitation of these technologies are licit by now. Such technologies are targets for competitors and copyists since no legal protection can be obtained by any party. Typically, spare parts with high margins, which already ran out of patent protection are targets of imitations and can only be protected by factual means.

Usage of the Positive Connotation Without IPR Infringement A well-known brand and its positive connotation are attractive for imitators. To take advantage of this positive reputation, competitors in China are trying to legally exploit the image of a well-known brand. The usage of similar marketing strategies and a similar corporate identity such as similar trademarks or the usage of similar color combinations, fonts and packaging materials are only some means experienced. Since such actions are deliberately planned within the legal boundaries and the proof that similarities can mislead the customers is not necessarily explicit, make it particularly difficult to counteract such activities.

Competitive Pricing and Second Mover Advantages Competitive pricing strategies can lead to a decrease in market prices and reduced margins. The licit or illicit imitation of products allows a reduction of research and development costs as well as fast market entries; often illicit distribution chains are exploited. Compared to the original manufacturer, the imitator has a second mover advantage, which allows him to exploit a competitive pricing strategy.

The legal threats and factual challenges are manifold and vary in their characteristics and creativity. Although legal and factual challenges are not exclusive to China, these cases have shown that China-specific differences such as IP awareness, language, and cultural differences as well as weak legal enforcement exist. The risk of occurrence and its impact can be higher compared to countries with stronger appropriability regimes and more difficult to counteract.

The different IP threats usually result in three **damages**:

1. Loss of revenue due to decreasing demand (of products, spare parts, services), absence of royalties/licensing fees, loss in market shares due to decreasing number of customers, price war due to dumping prices
2. Reputational damages (misuse of brand names and reputation)
3. Liability damages

In China the impudence and creativity of infringements show that the extent of IP infringements and challenges remains a difficult task to counteract. We have collected our research data from dozens of western firms that entered the Chinese market within the last three decades. We collected successful practices and out of these developed the 'IP protection star'. The next chapter presents the 'IP protection star' in detail.

The IP Protection Star

4

Firms need a holistic piracy intelligence and IP protection to profit from innovation. They also need to develop strong brands and to manage technology-intensive processes in China. The ingenious ways of legal-illicit, legal-licit and other factual IP threats in China demand a creative protection management. The different facets of legal and factual protection means need an overall perspective of directly or indirectly protecting IP.

The IP protection star addresses five dimensions, namely *legal-*, *technology-*, *business-*, *market-*, and *human-driven* IP protection (see Fig. 4.1). Some of the firms we have observed solely focused on legal aspects and thus lost their IP by means of personal fluctuation. Some firms are excellent in technology protection but failed in the market because fake products with low quality damaged their brands. Some firms do focus on the business and the legal part, but were imitated by firms that received technical information from key suppliers.

All five dimensions of the IP protection star are equally important. The star helps managers to avoid white spots and to establish a holistic view on their IP management—a chain is only as strong as its weakest link.

Within each IP protection dimension different characteristics of application are identified and described. The examples reveal that the dimensions are not mutually exclusive but rather complementary in their characteristics.

4.1 Legal-driven Protection

4.1.1 Legal-Proprietary Protection

Applying IPRs in China The legal-proprietary protection comprises the entire scope of intellectual property rights and their application. The IP system in China is a triple system with three inter-related national powers: (1) legislative guidance, (2) administrative control and (3) judicial enforcement (Yang 2003b). *Legislative guidance* comprises the legislature authorities and the laws such as patent, trademark, copyright and anti-unfair competition law. *The administrative control* represents the

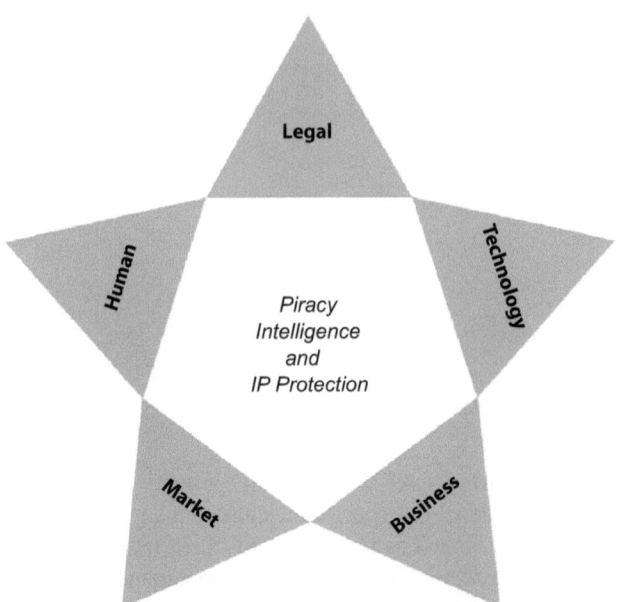

Fig. 4.1 The IP protection star

administrative organs, such as the SIPO, the Trademark Office and the State Copyright Administration. Their functions are the administrative procedures for patent applications, examinations, approval and protection. *The judicial enforcement* comprises the different courts and their functions in dealing with IP disputes. The current Chinese patent law and the Implementing Regulations of the Patent Law protect three kinds of patents:

1. Invention patents
2. Utility models
3. Industrial design patents

The Patent Application Procedure in China Since China has been a member state of the Patent Cooperation Treaty (PCT) since October 1993, the PCT application is an eloquent way to apply patents in China. Under the PCT, the patent application can be delayed to file the application in individual national patent offices for up to 30 months after the first priority filing date. For an extra fee, the duration of 30 months before entering the so-called national phase can be prolonged by a 2-month grace period. Thus, the PCT application enables the delay of the prosecution costs for filing applications in multiple countries. Furthermore, the prosecution costs are consolidated due to the single application format and the language advantage. It also renders the applicant a preliminary feedback in terms of the patentability of the invention and provides the opportunity to present arguments for patentability, to amend claims, and to strengthen the application prior to the filing at the national patent offices. This procedure streamlines the process of filing a patent application in multiple countries.

In China (like in most other PCT member countries), the national phase entry has to be entered by the end of the 30-month period after the first priority date of the application. The applicant has to file an application with each selected national patent office and pay the associated national filing fees. The patent application documents must be submitted in the Chinese language. A late filing for the Chinese translation of specifications and claims is not accepted and the patent applicant has to submit the documents as required for the type of patent (invention patent, utility model, design patent) concerned. For a PCT Chinese national phase application the following requirements have to be fulfilled:[1]

1. One copy of the PCT international publication including the cover sheet (with information of applicant, inventor, priority, etc.), specifications (with description, claims, abstract and drawings) and the international search report.
2. One copy of the preliminary examination report and annex.
3. Amendments made in the international phase needing to be effected in Chinese national phase application.
4. Any further amendments to be made at the time of national phase entry.
5. The name of the applicant in Chinese characters.
6. The name of the inventor(s) in Chinese characters (if they are Chinese).
7. Indication if the request for substantive examination should be filed simultaneously with the application.
8. Indication if the re-registration in Hong Kong should be filed.
9. Power of attorney, with the original signature or seal of the applicant. Form can be handed in within 3 months of the Chinese filing date.

If the applicant wishes to reduce the number of claims, they should be reduced before the international publication to reduce the costs since the number of claims is calculated based on the international publication specification. The amendments are allowed but should be filed as substitute pages. Prior art documents or lists of prior art only have to be submitted upon the request from the examiner, a voluntary submission is not necessary. If the applicant has a local site, the patent application can be submitted directly at the Patent Office. Or, if no local site exists, the applicant should appoint a designated, authorized patent agent. Due to technical complexity involved in a patent application, foreign applicants have to appoint designated agents, which have to be authorized by the SIPO. If a patent agent handles the application process, an authorization letter, namely power of attorney confirmation is required.

Preliminary Examination of the Patent Application   After the SIPO receives the invention patent application, the office verifies the conformity with the requirements and publishes the application after 18 months from the date of filing or respectively 30 months after the priority date. Upon request of the applicant, the intellectual property office may publish the application earlier (HKTDC 2005). The preliminary examination has to check the formality of the application, which are the completeness of the documents, the submission in a timely manner, and the correct

---

[1] See Appendix II for the requirements of non-PCT/national Chinese patent applications.

Case Example: Shanxi Museum

The Shanxi Museum in China has an overhanging front. The main building is shaped like "a Dipper on a Tripod" with four wings (see image below). In 2003 the front of natural stone was mounted onto the building; 50,000 metal anchors were used for the installation. However, the official anchor manufacturer (fischer) only delivered 5,000 anchors to the construction side. fischer was somewhat stunned that the construction was finished someday without a further purchase order.

© inmagine.com

It turned out that the contractor ordered 45,000 anchors from a Chinese company that had exactly copied fischer's anchor in form and size and even used the fischer logo on them (see image). However, the material used was not an exact copy—due to minor corrosion properties it is a long-term danger to the stability of the facade. Approaching different Chinese authorities was fruitless for a long time. Only after Germany's Federal Minister of Economics and Technology (Michael Glos) addressed this case to the Chinese Government, the copying was officially ascertained and fischer therefore released from guarantee obligations.

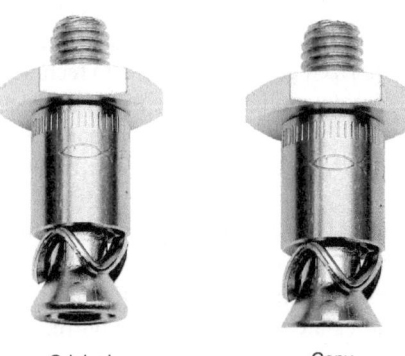

Original   Copy

payment of the application fees. During this period the patent applicant may correct any problems identified by the examiner. If the applicant cannot adjust any of the problems, the patent application is rejected. For a utility model or design patent application, only a preliminary examination determines if any rejections may apply. If no reasons for rejections occur, the intellectual property office grants the patent rights, publishes and registers them. Utility models and design patents come into effect on the publication date.

For a patent application to work the substantial examination has to be requested within 3 years from the date of filing (or from the date of priority if it is claimed) to the Patent Administration Department (PAD) of the SIPO. If the time limit cannot be met, the application is deemed to have been withdrawn. The substantial examination determines the patentability of the invention based on the three requirements *novelty*, *inventiveness* and *practical applicability*. If the examiner identifies any objections and the patentee cannot overcome these arguments, the application is rejected. If the applicant is not satisfied with this decision, the applicant can file a request stating his reasons for reexamination at the Patent Re-examination Board within 3 months from the rejection notification. The applicant may adjust the amendments to the application documents to conquer the deficits stated as rejection reasons to facilitate the Board to revoke the rejection made by the patent examiner. If the applicant receives an unfavorable decision again, he may institute legal proceedings at the People's Court within 3 months. If a patent has been declared invalid following the re-examination proceeding, it is deemed to not have ever existed.

If no reason for rejection is found after the substantive examination procedure, the decision to grant the patent right is made. The intellectual property office issues the certificate of the invention patent and publishes and registers the invention. The right comes into effect on the publication date.

Costs for Filing and Prosecution for a PCT Patent Application in China The entire costs for filing and prosecuting an invention patent application in China amounts to around RMB 14,000 Yuan. An additional charge of about RMB 3,000 Yuan may be charged if changes are made to the application after filing date, if deferments are requested, or if the specification and claims are longer than the specified page limits. Additionally, attorney fees have to be taken into account. In particular, the translation, depending on the number of claims, is a cost driver in the overall patenting costs. The costs for the filing and prosecution of utility models or design patents are considerably less and cost around RMB 4,000 Yuan.

Alternatively to the PCT procedure, the applicant may also apply a patent directly at the SIPO in China. Usually, a direct application only occurs when the invention was made in China and the inventor has to follow the rule of a priority application in China. The advantage of filing a national phase PCT application as opposed to filing a patent application at the national office directly is that the applicant can use the information acquired during the PCT international phase to strengthen the application upon entry into the national phase. In practice, the applicant can use the information derived from the written opinion and the international preliminary report on patentability to decide which claims to amend or eliminate prior to entry into the national phase to reduce costs (Schneiderman 2007).

Request for Invalidation of a Granted Patent After a patent is granted any individual or entity may claim that a granted patent right is not conform to the law and can request the Patent Re-examination Board to declare the patent right invalid (according to Art. 45, Chinese Patent Law). The required documents are the power of attorney letter, the title of the patent and the patent number, names, address, nationality of applicants and patentees and the reasons, the relevant material and the evidence for invalidity.

The patentee receives a copy of the request for invalidation and the relevant documents from the Patent Re-examination Board and should then present observations within a specified time limit. The patentee may amend the claims. Broadening the scope of the patent protection is not allowed. The Patent Re-examination Board examines the request for invalidation, makes a decision and notifies the requestor and the patentee. If the patent right is declared as invalid, the decision will be registered and announced by the PAD.

If one of the parties is not satisfied with the decision of declaring the patent right invalid or upholding the patent right, such party may institute legal proceedings at the People's Court within 3 months from the date of the notification. The Intermediate People's Court of the municipality of Beijing shall be the court of first instance, while the High People's Court of the municipality of Beijing shall be the court of second instance.

4.1.2   Protection by Legal Enforcement

Three different patent-related court proceedings exist in China: the *administrative*, the *civil* and the *criminal* proceeding. The rights holder usually starts with an investigation to determine the level and scope of infringement and also aims at the collection of evidence for the filing of a complaint. The investigations may render information about whether the alleged infringer manufactures and/or sells the patent protected product or process, the volume of infringing goods produced or sold and the possible location of such goods. This first investigation and collection of evidence may help the patentee to determine whether to proceed with a cease-and-desist letter, with raid actions and administrative procedures or with filing a civil lawsuit. In practice, criminal proceedings are usually less relevant for IP infringements, thus a two track system remains for enforcing IPRs.

Preconditions for Legal Enforcement The legitimacy to enforce one's rights demands the duly registration of IPRs in China. In case of a patent invention, the patent law provides a partial protection after publication against third parties exploiting the invention. In such a case, the patent applicant may ask the alleged infringer to pay royalty fees for the period between the publication and the registration of the invention patent. Only after the patent has been granted, the full judicial enforcement is possible. For utility model and design patents, the protection starts after the official registration. In the case of a utility model infringement, the plaintiff has to provide the administration with a search report since a substantive examination is not conducted during the granting process.

Since there is no meaningful discovery procedure in China, the plaintiffs have to collate and submit the evidence regarding the infringement and damages. As reported by several IP experts, the proof of evidence is a challenging and risky task in China. The evidence has to be presented in Chinese and must be notarized, which is time consuming and deadlines have to be met. The meticulous gathering of evidence, its amount and quality have a direct bearing on the chances of success in a patent litigation in China. In practice, the gathering of notarized evidence can be risky and dangerous considering the limited access to infringing parts, since the confrontation with counterfeiters of breaking the law may result in harmful conflicts. Also the timing with respect to translations and notarizations and the fast reactions and stashing of evidence by an alleged infringer impede the collecting of evidence in the required time. External support by agencies is common practice for raids and the collection of evidence at third parties' sites.

Two Track System: Administrative and Judicial Track to Enforce Patent Rights The administrative track is generally the faster and inexpensive solution. Depending on the sort of infringement (patent, trademark, or copyright), the administrative track renders different advantages or disadvantages, especially in case of a patent infringement. In case of a patent infringement, the patent holder may conduct an investigation; collect information (e.g., buy samples of infringing goods as direct evidence) and proof of evidence, which by all means needs to be provided in Chinese and needs notarial acknowledgement. He can then file a complaint at the local administrative office of the SIPO. The local administrative office has the authority to mediate and order the patent infringer to cease infringement. They may also seize and destroy the infringing goods or equipment used in manufacturing the infringing product. Whenever any infringement dispute relates to a process patent the manufacturing of a new product, the party, which manufactures the identical product, has to furnish proof to show that the process used in the manufacturing of its product is different from the patented process. In case of an infringement related to a utility model, the patentee has to furnish a search report made by the patent administration department under the State Council.[2] Moreover, if a party passes off a patent of another party as its own, the party will (in addition to bearing his civil liability according to law) be ordered by the administrative authority for patent affairs to amend his act and the order will be announced. In such a case, the illegal earnings will be confiscated and, in addition, the party may be imposed a fine of a maximum of three times his illegal earnings and, if there were no illegal earnings, a fine that is limited to the amount of RMB 1,000,000 Yuan. Normally, the procedure will take a relatively short period of time and no fees apply for administrative protection (Cheong 2006). However, the administrative track does not render any

---

[2] For more details see Chap. VII, Art. 57, Patent Law of PRC.

compensation, which is the case for the juridical track. The administrative track for patent infringement is mainly recommended if the infringement is obvious.

Challenging is the fact that the administrative office may decline to take action due to the complexity of the case or insufficient technical and legal support within the administrative office. Then the legal proceeding can be instituted. The juridical track may also be instituted in case the administrative track fails in mediation.[3] Furthermore, the infringer can appeal within 15 days from the order at the People's Court if not satisfied with it. In such a case, the advantages of the administrative track perish. Finally, if the patent holder has started the administrative proceeding and later files a lawsuit alleging the same infringement case, the fact of simultaneous judicial proceedings can cause the dismissal of the administrative proceedings.

Thus, deciding for the juridical track is often the case if the infringed party anticipates high compensation damages. This track applies to the Civil Procedural Law since litigation falls under civil litigation. The jurisdiction of the court has to be in place where the infringer has his/her domicile or where the infringement takes place.

Strategic Selection of Place of Venue The location where the infringement takes place includes the place where the infringing products are manufactured, used, or sold. According to Art. 29 in the Civil Procedure Law (1992), the place of venue is the location where the infringement takes place or where the defendant has his domicile and the lawsuit should take place under the jurisdiction of the People's Court.

In an interpretation by the Supreme People's Court, the place where the infringement takes place includes not only the place where the infringing act actually takes place but also the location where the consequence of the infringement occurs.[4] In practice, this interpretation has been debated although the Supreme People's Court has provided further guidelines, e.g., the place where the consequence of infringement occurs should be the place where the direct damage due to the infringing act has occurred but not necessarily the place where the plaintiff has suffered damage due to the infringement (Cheong 2006). Thus, the place of venue can play a strategic role for subsequent legal proceedings. Since the place of venue determines the administrative authority in charge, it can be selected, so that as little as possible relations might exist between the implementing bodies and the infringer. The selection should consider the local protectionism, the experience of the judges and the overall assessment of expected success in the case of litigation. The network and relations of an infringer need careful investigation before entering the administrative proceeding. The rights holder can either initiate the administrative proceeding, a civil proceeding or request for the issuance of an immediate injunction order.

As in any lawsuit, the proof of evidence is the key to success and a basic precondition to the litigation. If a party is infringed, one of the first actions is the collection of proof of evidence. The plaintiff has to submit evidence in order to initiate legal

---

[3] For more details see Civil Procedural Law of PRC.

[4] For more details see Supreme People's Court Interpretation, Sect. 5 and 6, 22 June 2001.

proceedings. The laws and regulations do not include any discovery procedures. Thus, the collection of prima facie evidence can be frustrating for the patent holder although he clearly knows of the fraudulent infringement activities. One means for collecting proof of evidence is the purchase of the infringing product. Although such evidence might not be admissible in the trial because of lack of authenticity, it is usually sufficient as prima facie evidence for the court to accept the case.

In case that the alleged infringer immediately stops the infringement and complies with the infringement complaint, the plaintiff might run the risk of bearing the burden of the procedure costs and of the refundable attorney's costs for the alleged infringer. Once the court establishes the case, the patent holder can then request the court for further evidence collection, if the patent holder cannot collect such evidence because of objective reasons; e.g., the third party such as a bank might refuse to disclose evidence until it is ordered to do so by the court, or if the court deems such further evidence necessary for the hearing (Cheong 2006). Furthermore, if the evidence may be destroyed or lost, or if it is difficult to obtain it afterwards, the patent holder can request the court for an order of evidence preservation with or without notice to the alleged infringer depending on the type of evidence to be preserved.[5]

Generally, the likelihood for a notice is low for raids but most other situations request a notice in advance. The court has to decide on the request within 2 days. In the case that the court issues the order for evidence preservation, it can search the alleged infringer's factory building, including warehouses, sample rooms, workshops as well as offices. The sizeable objects include samples, sales contracts, invoices, and account books, which will be used as evidence in the trial. In contrast to the pre-suit injunction, the patent holder does not have to pay a fee or if he does, it generally is a relatively low amount. The next burden is the decision whether or not the court accepts the evidence. There are very strict levels of the authenticity of the evidence and the court requires the originals of all documents or materials, which often require notarization and foreign evidence may additionally need to be legalized. In case that the infringing product relates to a process claim, the party, which manufactures the identical product, has to furnish proof to show that the process used in the manufacture is different from the patented process.[6]

Cease-and-Desist Letter In a first step, the rights holder obtains knowledge of activities that infringe a patent right and of the infringing party to determine the impact of the fraudulent activities. In a standard situation of a low impact infringement, the rights holder can either send an inquiry letter or a cease and desist or warning letter to the alleged infringer. In the inquiry letter, the rights holder asks for an explanation why the alleged infringer feels entitled to make use of the patented invention. In a cease and desist or warning letter, he requests from the alleged infringer to cease and desist the use and production of any infringing goods already produced or distributed. The letter may also seek an undertaking not to infringe in the future and an agreed amount of liquidated damages if breached, and compen-

---

[5] See Sect. 65, Civil Procedural Law.
[6] For more details see Patent Law of the PRC, Chap. IX, Art. 57.

sation for past infringement. Such a letter may also indicate that in case of non-compliance with the request, the rights will be claimed against the alleged infringer on the basis of the IPR in question. To ensure that the rights holder can manage the pace of action, such a letter should include a response term for the alleged infringer. If the alleged infringer does not reply within that time, the rights holder should issue a second letter accompanied by follow-up phone calls. By means of such a notification and request, the rights holder obtains proof that at least from that time onwards; the alleged infringer knew that the rights holder considered his activities as patent infringement. Another goal is deterrence and to exude an offensive IP protection.

In case of a settlement arrangement, the rights holder can usually attain the immediate stop of infringement and depending on the case pursue a monetary compensation. He may also request the publication of an apology advertisement in a major metropolitan newspaper. In case the involved parties are not willing to resolve the conflict by correspondence, the rights holder begins to prepare evidence, which could be used to file a patent infringement complaint against the alleged infringer at the court of venue. Legal proceedings can also be started by the rights holder without a preliminary correspondence and without a preliminary cease and desist or warning letter to the alleged infringer.

Administrative Procedure and Raids To enforce IPRs the nature of the IPRs infringed and the complexity of the case as well as the evidence provided determine which agency to use. *The technical supervision bureau, the administration for industry and commerce, the patent administration office* and *the copyright office* are all designated separately or jointly to enforce IPRs. One means of these agencies is the completion of inspections of alleged infringers. Some are empowered to conduct raids and to actually seize infringing goods, impose fines or gather further proof of evidence. The administrative procedure is an inexpensive and fast alternative to the more costly and lengthy court proceedings and allows the omission of an infringement and the deterrence of alleged infringers. Thus, many rights holders prefer the administrative adjudication of patent infringement because the investigations may occur soon after the filing of the complaint. Other benefits are that the rights holder may be able to participate in the investigation and the time required for determining whether an infringement has occurred can be shorter than a legal procedure in court.

The administrative procedure starts with a request for administrative investigation at the local SIPO office where the infringing activity takes place or is believed to be taking place. For the request, the rights holder should carefully prepare the essential documents, that is the documentation to proof registration of proprietary rights and payment of annual fees, power of attorney as well as the proof of evidence of an infringement (von Welser and González 2007). Once presented to the authorities, they decide within a few hours if and which measures will be initiated. The measures, which can be initiated by the authorities, are:

- Injunctions
- Mediation upon the request of the parties
- Cease-and-desist orders

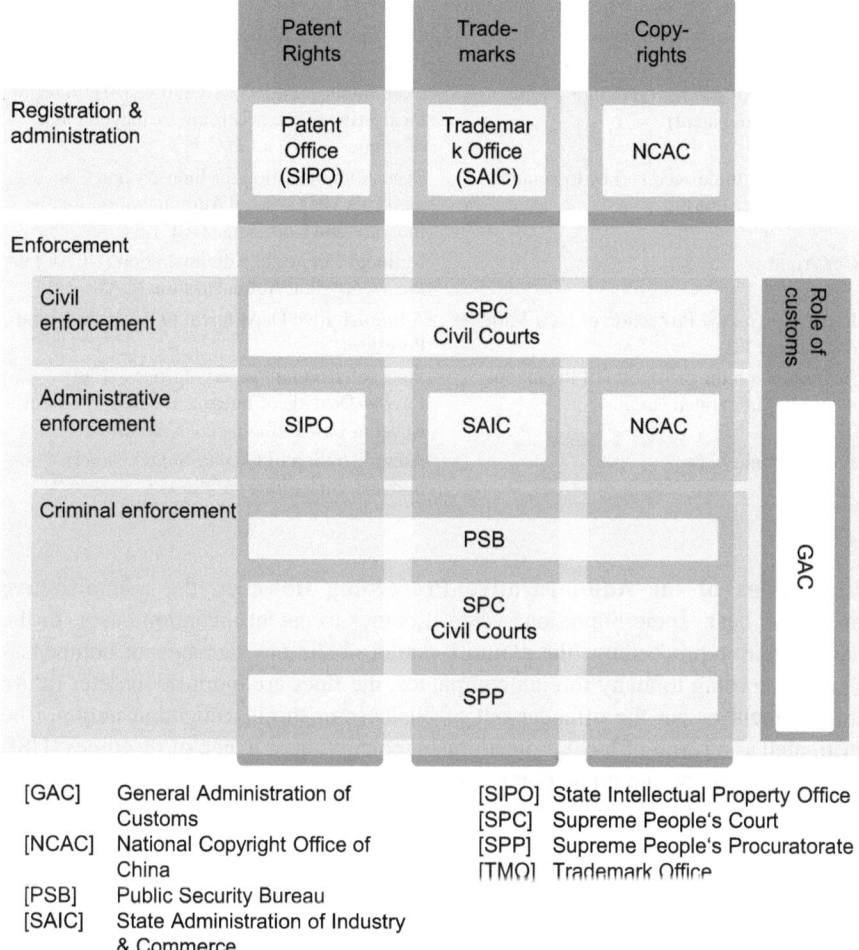

Fig. 4.2 Legal bodies for registration and enforcement of IP in China. (Source: Misonne and Ranjard (2006))

- Confiscation of illegal earnings
- Fines of not more than three times of the illegal earnings or (if there are no illegal earnings) no more than RMB 1,000,000 Yuan.

The imposed administrative actions or fines may be appealed to the People's Intermediate Courts. If the infringement constitutes a crime, the infringer shall be prosecuted for his criminal liability. An overview of the legal bodies for the registration as well as the enforcement is presented in Fig. 4.2.

Depending on the relevant IPR of an infringement, the corresponding administrative authorities vary. For invention, utility model or design patent infringements the SIPO or local SIPO office depending on the alleged place of venue are the responsible bodies (Table 4.1).

Table 4.1 Overview of different administrative authorities according IPR infringements

Relevant IPR of an infringement	Corresponding administrative authority in China
Three kinds of patents (invention, utility model, design patent)	State Intellectual Property Office (SIPO) or local SIPO office according to alleged place of venue
Trademarks, trade secrets (law to counter unfair competition)	State Administration for Industry and Commerce (SAIC) or local Administration for Industry and Commerce (AIC)
Copyright	National Copyright Administration (NCAC) or local Copyright Administration (CA)
Regulation on the Protection of New Varieties of Plants (PVP)	Administrative Department of Agriculture and Forestry
Regulation on the protection of the layout designs of integrated circuits	State Intellectual Property Office (SIPO): Layout-Designs of Integrated Circuits Administration Law Enforcement Committee
Product quality law	Administration of Quality Supervision Inspection and Quarantine (AQSIQ)

Limitations of the Administrative Proceeding However, the administrative procedure bears some limitations when it comes to patent invention cases. In the administrative proceedings the plaintiff cannot obtain any damages or compensations. According to many foreign companies, the fines are too little to deter future infringement or put the offender out of business or the investigation may not be instigated as a result of local protectionism, corruption or a lack of resources (USE 2010). Also, an interim injunction is not possible for an ad hoc omission. According to the regulations of the patent infringement administrative proceedings, the alleged infringer has to be notified in advance of the pending action. The administrative authorities may also decline a patent infringement case if it is technically too complex. In that case, the rights holder can proceed with a civil infringement procedure. In any case, if an administrative enforcement case is initiated, the validity of the patent may be appealed to the local court having jurisdiction over that administrative agency. Thus, it can bear the risk of an invalidation of one's own rights.

The drawback of the administrative proceeding is the fact that no damages or compensations are enforceable. In case of a patent infringement an interim injunction is not possible without a court ruling. Furthermore, the administrative body may decline the processing of the issue due to the technical complexity of the case.

Although the administrative proceeding offers means to protect and enforce proprietary rights, in practice some unforeseeable limitations may occur. Although the administrative authorities make arrangements to cease and desist an infringement, the successful execution may fail due to local protectionism or corruption. Particularly in rural regions the power and protection of local employers might influence the administrative enforcement. Individuals, who are empowered to enforce the administrative actions locally, are often also involved in local committees, which lead to inscrutable antagonism. In a survey conducted by the European Commission, the three main factors influencing the identification of infringers are the *local protec-*

tionism, the *protection of the original producer by criminal organizations* and the *restrictive nature of investigative work* which foreign right holders are allowed to carry out (EU Com 2006). Thus, the relational networks of the infringer need to be investigated since they influence the enforcement of the compiled administrative arrangements.

The obscurity and lack of publication of former cases and administrative rulings impede the evaluation of previous incidents. There are also no continuous publications on the fines and administrative actions that were arranged. Furthermore, the affected companies report that fines have been assessed but could factually not be enforced. The following table summarizes the advantages and limitations of the administrative proceeding (Table 4.2):

IP Enforcement by Civil Litigation Despite a critical public view in China and the skeptical opinion of foreign business communities with regards to China's legal system in general, IP matters enjoy the highest priority in Chinese courts. Moreover, legal staff members—including the relevant judges—are increasingly coached in subjects concerning intellectual property. Both the Intermediate People's Court—which is the appropriate court for IP infringement in the respective region where the action is filed—and the High People's Courts have significant experience with IP trials (Hane 2008).

As soon as a patent rights holder notices an infringement, a period of 2 years is granted to legally enforce the violation. Preliminary injunction can serve the patentee in the event of reasonable evidence of infringement or imminent infringement. Damages are determined according to the losses suffered by the patentee or the profits gained by the infringer from the infringement. In the event that none of these methods are appropriate to quantify the damage, the patentee may optionally claim a reasonable royalty for the respective patent (USE 2010).

Initiating Civil Litigation A written complaint to a People's Court officially initializes the lawsuit. The letter has to contain the following elements:

- The complainant's name and address
- The name and position of its legal representative (if applicable)
- The nature of the claim and the facts upon which the claim is based
- Evidence and sources of the evidence
- The names and addresses of witnesses

Evidentiary Requirements and Burden of Proof As opposed to a concept of new inventions such as of the United States, in China each party is accountable to prove their claims of originality. If the People's Court for a case deems specific evidence relevant, the court itself may assemble it. The same procedure applies in the case that a party is not capable of providing the proof due to objective circumstances. All relevant facts have to be presented in court. The court reserves the right to decide upon authenticity and validity of the evidence.

Table 4.2 Summary of administrative, civil and criminal proceedings and advantages and limitations

Administrative proceedings

Legal body	Advantages	Limitations
Different administrative authorities, depending on the type of infringed rights	Inexpensive and fast	No damages or compensations possible
	Local	Patent infringements can be declined due to technical complexity
	Expert agency (nearly every IP right has an administrative agency)	Authorities only make arrangements if infringement is clear-cut and obvious
	Admin. agency may have other enforcement capabilities	Decisions may be influenced of local protectionism and corruption
	No lawyer needed	Arrangements of patent authorities can only be enforced by the order of court
	May be able to obtain an injunction	Local protectionism
	Possibility of fast actions to get access and confiscate goods (e.g., raids, searches)	Penalties usually *non-deterrent*
	Option to obtain evidence for subsequent legal proceedings	Non-transparent
	Omission and fines can be asserted	Difficult to transfer to criminal prosecution
		Limited geographic jurisdiction

Civil proceeding

Legal body	Advantages	Limitations
Different administrative authorities, depending on the type of infringed rights	Damages available	High costs
	Specialized judiciary who may be trained in IPR	Low and limited damage rewards
	Injunctive remedies	Lack of independence of judiciary in many jurisdictions
	Rights to appeal	Difficulties in collecting damages, if rewarded
	Nationwide jurisdiction	

Criminal proceeding

Legal body	Advantages	Limitations
Different administrative authorities, depending on the type of infringed rights	Deterrent damages possible (fine and imprisonment)	Time consuming
	Possibility of civil damages in addition to criminal punishment	Difficulty in case of business relationships to opponent

Available Remedies Potential options of IP lawsuit settlement include:

- An order to cease infringing activities
- Elimination of the effects of the infringement
- Issuance of a public apology, confiscation of unlawful gain or infringing products and assets used in furtherance of the infringement
- Compensation for damages suffered, as well as
- Any combination of the above

To prevent patent or IPR infringement, civil litigation is used in an increasingly manner in China. Instead of legal adjudgement, civil action provides a good alternative and is being applied by Chinese legal authorities more and more often. Nevertheless, significant litigation costs in correspondence to a low level of compensation need to be considered.

Criminal Prosecution IP infringements that show a significant level of violation can be prosecuted criminally after being re-assigned from other authorities, for example Chinese customs. Initially, the Public Security Bureau is involved; the case may be afterwards transferred on to the Supreme People's Procuratorate.

Numerous rights holders from countries outside of China criticize the unclear criteria with regards to a criminal case. Moreover, they claim that those criteria allow a wide spectrum of legal judgment, and entry barriers for criminal prosecution are set too high. Alongside of fines and compensation payments, patent infringement can result in 7 years of custodial sentence. However, the majority of IP violation cases are resolved through the administrative system.

The Role of Prior Art: Defense for Infringement or Ground for Invalidation
Defendants can initiate an invalidation procedure at the Patent Re-examination Board when they are accused of patent infringement. Other reasons for invalidation can be the expiration of the patent or the patentee's abandonment of the patent.[7]

Prior art[8] can be a defense to infringement claims if the subject matter of the patent would have been obvious at the time the patent application was filed. However, the use of prior art in infringement cases is twofold: In relation to utility models and design patents the courts are more likely to accept prior art as a defense for infringement, whereas in relation to invention patents the courts may disregard the prior art defense unless the patent office declares the patents thereof invalid through an invalidation proceeding (Kluwer 2005). The reason is the concern of the judges

---

[7] According to Arts. 88 and 91 of Opinions of the Beijing High People's Court on Several Issues Relating to Patent Infringement Establishment (for trial implementation).

[8] The term 'prior art' is defined in Rule 30 of the Implementing Rules of the Patent Law as "any technology which has been publicly disclosed in publications in the country or abroad, or has been publicly used or made known to the public by any other means in the country, before the date of filing (or the priority date where priority is claimed) is prior art". In Art. 22 of the Patent Law (2000) that term and definition is used.

Case Example: Hilti

Hilti offers innovative products, systems and services to construction professionals. The company is active in more than 120 countries with some 22'000 employees. The Hilti brand is well known in the construction industry and its reputation has influenced the entire drilling and demolition segment, professionals refer to such tools by "Please give me the *Hilti*". In China, Hilti is called "XILIDE" which was selected due to its positive connotation, meaning "to have luck, get profit". According to brand experts, the Hilti brand has a value of more than one billion Swiss Francs.

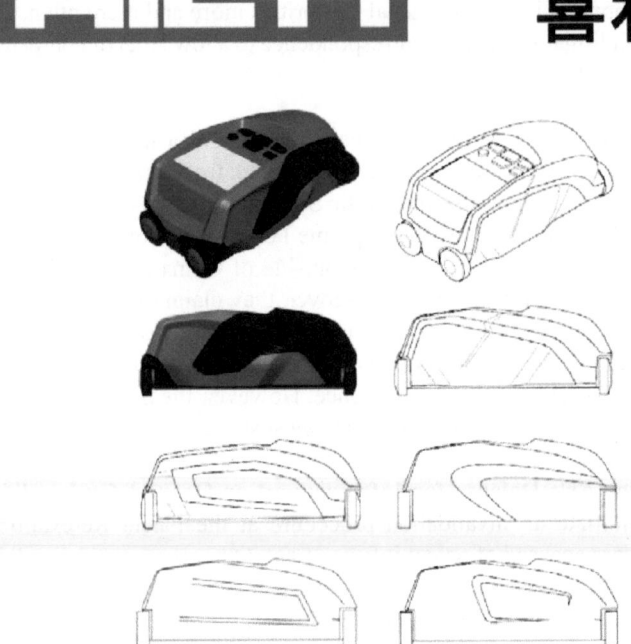

The Hilti innovation PS 38 Mulitdetector is a highly versatile detector for locating metal, electrical conduits, wood and plastic objects in base material. It allows the construction professional to identify where he may drill a hole. The tool detects the location of objects, its approximate embedment depth and type of material. The ergonomic design is the first of its kind. The product is ruggedly built, shock, dust and water resistant and easy to steadily run over the material due to its wheels. Hilti protects the customer's value proposition in multiple ways: complementary to patent and trademark protection, the ergonomic design is also protected by a design patent. The design patent protects the Hilti design as well as the mimic or potential alternative designs.

to disregard the patent claims of a valid invention patent as compared to utility and design patents that are not subject to substantive examination.

The *compensation income for the damage* caused by the infringement shall be assessed (Art. 60, Patent Law of the PRC):

1. On the basis of the losses suffered by the patentee or
2. The profits which the infringer has earned through the infringement or
3. If it is difficult to determine the losses, which the patentee has suffered or the profits, which the infringer has earned, the amount may be assessed by reference to the appropriate multiple of the amount of the exploitation fee of that patent under a contractual license.

4.1.3   Legal-Contractual Protection Means

Legal-contractual protection means incorporating all contractual agreements which include IP-related issues. Such agreements usually either concern internal agreements with employees or external agreements with suppliers and partners. In both cases the protection of proprietary know-how should be included into the contract. With suppliers and partners, the degree to which the protection has to be determined depends on the cooperation content. In case of mutual development activities, the foreground and background know-how as well as the residual information, which remains in the heads of employees, should be determined contractually in the beginning of cooperation (Bader 2005). Particularly in R&D cooperation, the contract negotiations and the clear agreement on IP rights is essential for a potential enforcement and its own freedom of action. The latter can be endangered if no clear agreement takes place. According to the law, the results of cooperation belong to both parties and thus, in case of licensing or usage of IPRs one party may block the other of the use or may claim royalties.

The internal agreements are usually employment contracts or additional confidentiality contracts for employees. The agreement should include non-disclosure terms as well as non-competitive clauses. In China the non-competitive clause is essential since a main challenge is the knowledge drainage after an employment. In all four case studies the interviewees have anticipated the perceived risk of knowledge drainage to be higher in China than in western countries. Although an agreement is no guarantee for fraudulent know-how exploitation, the creation of IP awareness and the potential threat of legal enforcement are deterrence factors. The following example shows the written IP regulations as part of the employment contract of a European MNE. It includes:

1. General provisions
2. Protection of Intellectual Property of the company
3. IP-related duties of the employee in English as well as in Chinese, which are part of the employment contract at the Chinese subsidiary

Example:
IP Policy Established in China Including R&D Employees' Duties

1. **General Provisions**

   This IP policy is an integral part of all employment contracts for all employees. The obligations of this IP policy shall apply to the time of employment with the company, and thereafter, if related to employee's prior employment with the company and resulting from the character of the obligation. Therefore, the company may adopt more detailed IP regulations for those employees, which by nature of their duties have a special need. The company may modify or amend this IP policy from time to time if deemed necessary by the company.

2. **Protection of Inventions and other Intellectual Property of the Company**

   2.1. The company is determined to protect its intellectual property rights related to inventions of its employees as well as related to other intellectual property, whether it is patentable or not (hereinafter referred to as 'Intellectual Property'). Such intellectual property includes but is not limited to discoveries, inventions, creations, improvements, trade secrets, know-how, patents, utility models, design protections, and copyrights.

   2.2. All decisions regarding filing, foreign-filing, prosecution, defense, enforcement, maintenance, abandonment of the intellectual property shall be at the sole discretion of the company. The company is not obliged to inform the employee about any of its decision or obtain his/her consent to said decisions.

   2.3. All intellectual property rights related to all intellectual property, which the employee obtains during the employment, throughout the world shall be vested in the company. Furthermore, all intellectual property rights related to all patentable inventions, which the employee obtains within one year after his employment, shall be vested in the company, if related to the employee's work as described. The employee shall assign to the company all his/her intellectual property rights related to such intellectual property or inventions. The company may assign its rights under this IP policy to any party in interest: (As an exception, if the company cannot use any part of the intellectual property in its business , the employee shall be entitled to all intellectual property rights related there to, provided that with regard to patentable inventions the company has released its rights in writing.)

   2.4. The employee shall assist the company during his/her employment as well as thereafter at any time to obtain, defend, and enforce its intellectual property rights.

3. **Duties of the Employee**

   3.1. The employee shall comply with the following duties:

   a) During the employment: The employee shall promptly disclose to the company fully and in writing all intellectual property he/she obtains during his/her employment. The employee shall maintain adequate written records of all intellectual property. Those records shall be the sole property of the company and the employee shall make them available to the company.

   b) Within one year after the employment: The employee shall promptly disclose to the company fully and in writing any patentable invention he/she makes, and all intellectual property rights related to such inventions shall be vested with the company, in the event such inventions are related to the employee's work as described in Section 2.3 (d) above.

   c) Within two years after the employment: The employee shall promptly disclose to the company each patent application, in which he/she is named as inventor, and which is filed within two years after the termination of the employment and he/she shall explain to the company why such patent application did not result from his/her former employment by the company.

   3.2. Information an about conflict of interest: The employee shall immediately advise the company in writing about any potential conflict-of-interest if the company intends to entrust work to the employee, which relates to a field where the employee has done work for a previous employer not more than one year before and which may result in an invention. (Source: Extract of an IP policy of a European MNE with R&D facilities in China (2007).

In addition to the external and internal agreements, the publication of approval for any external communication should be a working principle to reduce the risk of knowledge distribution outside the company.

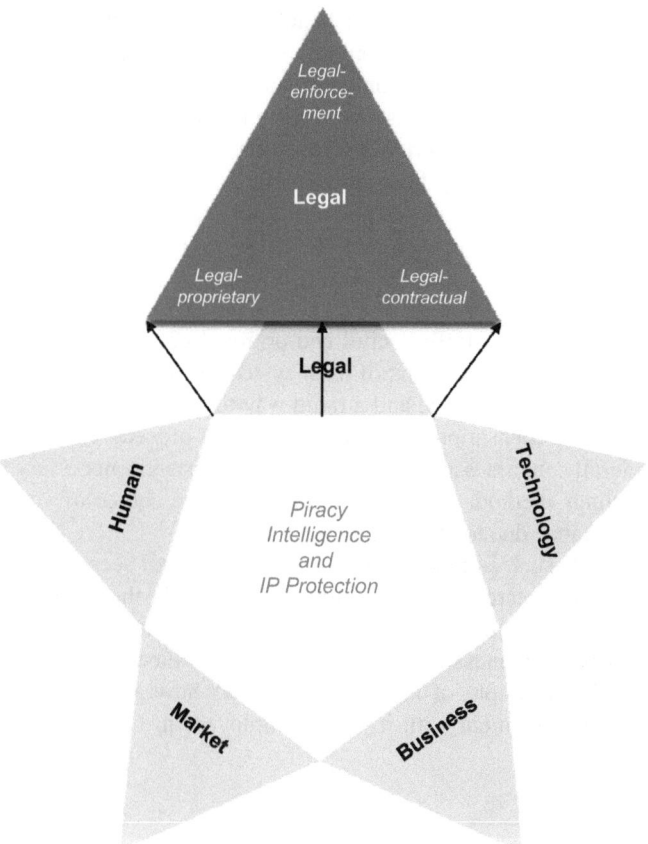

Fig. 4.3 Legally-driven protection means

Legal-enforcement
- Enforcement of IP rights
- Enforcement of trade secret

Legal-proprietary
- Application of invention patents, utility models and design model patents
- Trade secrets

Legal-contractual
- Secrecy agreements with employees
- Non-competitive clauses in employee contracts
- Non-disclosure agreements with suppliers and partners

4.1.4   Summary Legally-driven Protection

The application of IPRs represents the baseline protection required in China. Also contractual agreements with employees and third parties who receive or develop confidential information are also grounds for protection of IP by legal means. If no IPRs exist, enforcement against infringers is impossible; and no legal deterrence strategy can be applied. The application of design model patents may help to prove an obvious infringement and thus facilitate the enforcement of infringements in

China. The protection of the customer benefits generates value through IPR. Proprietary rights allow for the protection of the value proposition and thus protect the differentiation of the firm against competition. If enforcement actions are necessary, the majority of enforcement cases start with a request to halt the infringement or with an administrative procedure to obtain evidence and to stop the infringements. The civil enforcement is only favorable if notarized evidence is secured and compensation damages are high and the probability of success is strong. The advantages and disadvantages of the three possible methods of enforcement in China are presented in Fig. 4.3.

Next to the proprietary protection, legal enforcement is based on a tripartite system that includes administrative, civil and criminal proceedings. When weighing up the advantages and limitations of each system, it seems that an administrative proceeding is common practice and a rapid way to stop imitation. However, in the event that high compensations are expected, a civil procedure might be more adequate. Generally, an assessment on a case-by-case basis is necessary to select the right proceedings method. It should be noted that the choice of venue can be of strategic importance due to local protectionism.

With contracts such as non-disclosure agreements and secrecy agreements, the disclosure of trade secrets can be prohibited. Although, the enforcement of such contracts may be difficult, practitioners recommended the use of such agreements to at least create awareness that the issue is of relevance to the firm and for any cooperation. The example of an IP policy shows how a multinational firm has implemented an agreement with its R&D employees in China additionally to the employment contract.

4.2   Market-driven Protection

4.2.1   Market-Monitoring of IP-related Know-How

An IP rights holder needs an accurate understanding of the competitive market situation including (potential) infringers. Due to language and cultural differences as well as the vastness of China, the monitoring of the competitive market situation can be a major challenge for foreign firms. Often, the copying of products or the drainage of know-how by imitators is realized once imitations occur in the market. Technology and market intelligence measures help to identify the technology development in the market. On the one hand key technology fields of the company need an ongoing monitoring and scanning. This includes the monitoring and scanning of

- Industrial research activities
- Governmentally funded research programs
- Industrial expert working circles
- Activities of universities and research institutes and the corresponding industrial partnerships
- Conferences

On the other hand sales activities of competitors and suppliers have to be monitored. Regular audits of suppliers help to deter possible illicit production of proprietary technologies. Furthermore, the active search for IP infringing products, e.g., at trade fairs is helpful to identify infringers. Catalogue material should be monitored on a regular basis since some infringers even use the information and pictures of proprietary products for their merchandizing. In the manufacturing and capital-intensive goods industry, proprietary aspects are not visible but inherent in technical data. The analysis of written offers for public tenders may render important insights in possible infringements.

The monitoring of suppliers' activities is important to detect the fraudulent use of information, the over-production combined with complementary sales of proprietary components or products. In addition, the quality control of suppliers is essential to reduce the own subsequent rework. The integration of quality control can reduce efforts and costs and facilitate the control of suppliers. Many firms conduct unexpected audits at their suppliers' sites to ensure the licit production. Unannounced audits may have a negative impact on the trust building amongst the cooperation partners. The reason and goal of an unannounced audit versus a negative impact on trust building of a partnership need to be carefully assessed upfront.

The documentation of IP infringements should accompany the market monitoring to identify infringements. Such a database is necessary to track the available data of IP threats and should include the relevant information such as:

- Details of the infringing product
- Locality of the infringement (location and destination if stopped in transit)
- Details about the involved parties (retailer, importer and original source)
- Markets supplied
- Method of distribution
- Reaction/enforcement action or reason for no action

The information can be gathered by staff members, private investigators/informants, attorneys, or in some instances by the administrative agencies. The systematic documentation allows the monitoring and transparency of IP threats in the long run.

4.2.2 Market-Relational Efforts to Enhance IP Awareness

Guanxi is embedded in the Chinese culture and relationships and is essential for doing business in China. Guanxi describes the basic dynamic in personalized networks of influence. The Chinese word is becoming more widely used instead of the two common translations into 'connections' and 'relationships', as neither of those terms sufficiently reflects the wide cultural implications that guanxi describes. Closely related concepts include that of *ganqing*, a measure which reflects the depth of feeling within an interpersonal relationship and *renqing*, the moral obligation to maintain the relationship, and the *idea of face*, meaning social status, property, prestige, or more realistically a combination of all three (Gebauer et al. 2008).

In a corporate environment, guanxi has an external as well as an internal dimension. When doing business in China, the external guanxi embraces the relationships to Chinese suppliers and partners. Such relations need personal engagement. Also the relationship to Chinese governmental authorities can influence the success and support in doing business and thus also in IP protection. In fact, establishing good relationships with official bodies and institutions that formally may had little to do with IPR protection, may be a lot more effective than the legal system when it comes to IPR protection (Keupp et al. 2009). The de facto power of these bodies makes them attractive for strong relations and social engagement. Through these relationships they might win the status of an old friend, so that these bodies treat the foreign firm as their protégé and actively pursue IPR infringements (Keupp et al. 2010). Many interviewed managers report that their companies sponsor regular meetings and workshops to share information and to establish and maintain strong relationships with the Chinese legislators and local government representatives. One manager even reported invitations to the local mayor for important company events, e.g., the opening ceremony for a new production facility.

Furthermore, multi-national firms such as ABB, Philips, and Siemens sponsor IP-related courses at universities and institutes. The reputation of the company as a socially supportive and innovative firm also helps to win the best talents in the market. Moreover, a corporate social responsibility helps to create thorough relationships within politics and social networks. Tight contacts with the local government officials are therefore often maintained by social engagement and contribution, both with and without regards to the own business and IP issues. In contrast, the relationship to custom officers mainly exists with regards to IP-related issues to control exports and imports. Although multi-national companies often realize such measures, the measures are not limited to large companies. One way to gain strength in numbers and efforts to enhance IP awareness and protection is the collaborative approach with other western firms. The coordination of lobbying activities and public education has been established in trade specific groups, such as the *International AntiCounterfeiting Coalition*, the *Business Software Alliance*, the *Partnership for Safe Medicines* and the *Motion Picture Association*.

The deep roots of guanxi and the importance of relationships show that the traditional and preferred mean of dispute resolution in China consists less of a confrontational process such as consultation, mediation and arbitration. Compared to litigation, these means are less complex, help to maintain or repair relationships between parties and are more flexible and less costly (Bosworth and Yang 2000). Next to the external guanxi, the relationship with the own employees is vital to reduce know-how drainage.

4.2.3 Market-Educational Efforts

The quality of similar or counterfeit products is often minor to the original product and available for a lower price. In case the customer buys a counterfeit knowingly, the lower quality can lead to an understanding, which over time educates the customer that products for higher prices offer higher quality and are fulfilling demands more satisfactory. The low quality of counterfeits thus de facto discourages customers to buy such a product more than once if high quality is required. Also, the spreading of bad experiences with low quality products is of educational character for potential customers.

In companies with a high degree of technical know-how inherent in their products and with a good customer network, the sales and service staff are also educational trainers. They explain the advantages of the products and may know bad practice stories of alternative, lower quality solutions or counterfeits. The training of custom staff is essential if the main threats occur due to exported counterfeits. The goal is to identify and track back the counterfeits to fight the menace at its source.

Another educational aspect can be achieved by proactive publications of counterfeits. Stihl for instance openly publishes their active battle against product piracy. Counterfeits that have been captured are publicly denounced, including even television broadcasting. While in the political realm the press remains the voice of the government, this is no longer the case for trade and economic issues. Since journalists in China can be held personally responsible for what they write, they are cautious in their research. Personal relationships are therefore also important in public relations. It is common practice that companies have to pay a certain amount according to the media price lists for press releases. Foreign companies use the media for both, publishing their efforts and for underlining the credibility for a long-term engagement in China.

The education of customers is not limited to the products, it includes the corporate social responsibility efforts as well as the innovativeness of the firm. Local press agencies are often preferred compared to international PR agencies; although often better in strategic PR services, the local agencies have a high understanding of the local culture, language and practices as well as current topics under discussion. Those local agencies often have relationships with local politicians, authorities, as well as other journalists over many years. The public and official denunciation of counterfeit products are only one instrument to educate the customer in terms of an offensive IP enforcement. Hence, it is mainly used to deter potential producers of copies and demonstrate an offensive strategy against IP infringements and product piracy.

These indirect protection means deter imitators, educate customers and customs and make use of public relations and different communication channels. Due to the weak appropriability regime in China, the expectations of success in enforcement are often low for suing an infringer; the education about an offensive strategy is pursued instead. Another reason for the exploitation of such means is the lack of own proprietary IPRs and thus no option for legal defense.

4.2.4 Summary Market-driven Protection

Important for IP protection in China is the knowledge of existing and potential challenges and IP threats in the first place. Often companies realize the existence of imitations or IP theft only after they occur in high quantities in the market and when they may have already impacted their sales share. Market monitoring inside and outside of China is important to identify infringements at an early stage.

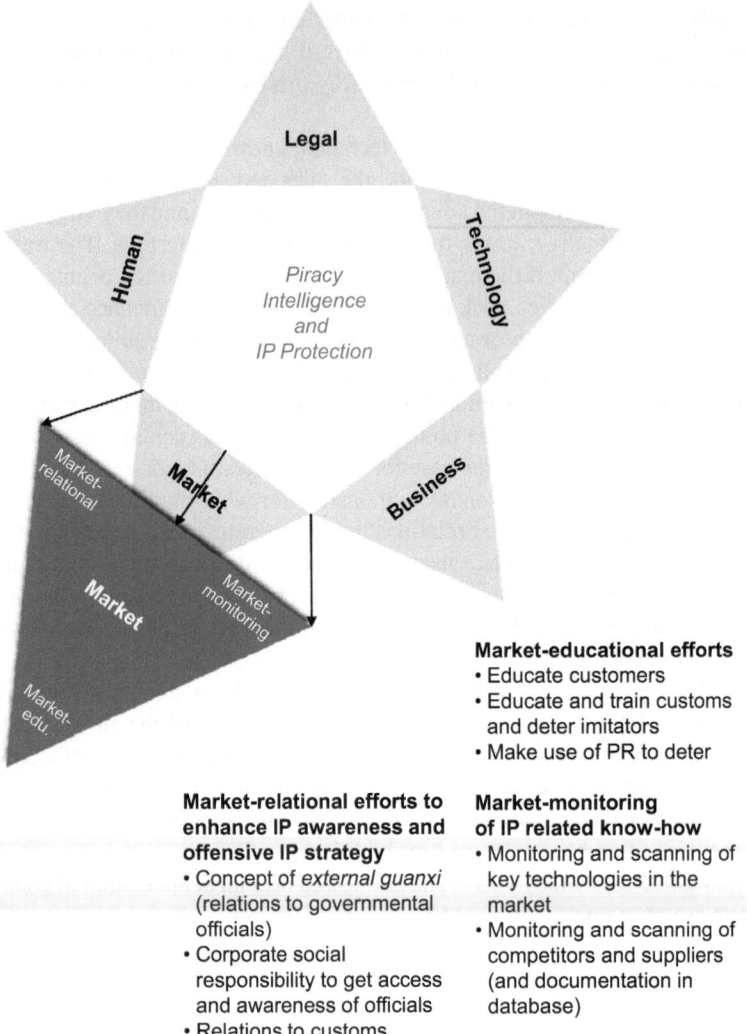

Fig. 4.4 Market-driven protection means

Market-educational efforts
• Educate customers
• Educate and train customs
  and deter imitators
• Make use of PR to deter

Market-relational efforts to enhance IP awareness and offensive IP strategy
• Concept of *external guanxi* (relations to governmental officials)
• Corporate social responsibility to get access and awareness of officials
• Relations to customs

Market-monitoring of IP related know-how
• Monitoring and scanning of key technologies in the market
• Monitoring and scanning of competitors and suppliers (and documentation in database)

Continuous monitoring and scanning of IP and key technology developments are one way to reduce the risk of a sudden rise of imitations.

In China, the concept of external guanxi should be exploited to maintain good relationships with governmental authorities. Corporate social responsibility activities are a method to access governmental communities and to underpin the long-term ambition of the firm. The long-term investment in local job security is important to build up supportive relationships with local officials (Fig. 4.4).

The education of customers is a method of factual protection that reduces the risk of a customer unknowingly buying an imitation. Education on the entire product offering and its long-term benefits, may also help gain new and retain existing customers. The education and training of customs about the product portfolio, the IP-protected features and the types of infringing products can help to identify and stop infringing goods before they are exported.

4.3 Human-driven Protection

China has a huge potential for competitive engineers and scientists. This is also due to their tough selection. Out of 1.3 billion people only about 250 million joined primary schools. Out of them only 5 million make it to a university. The selection is much tougher than in most western countries. Interesting is also the attitude of these children: A representative survey of the Shanghai magazine 'Attractive Power' has shown that 53 % of all 12–18 year old kids have 'a successful career' as their highest goal in life. Forty one percent of all 12–18 year old children have the goal to become a millionaire. This attitude, which one finds less in western peer groups, reflects the Chinese hunger and willingness to striving for wealth and progress. Loyalty towards an employer is far less important than the individual career.

Thus, major challenges are the protection of know-how in terms of drainage by current and/or former employees as well as cooperating partners' employees. Different aspects of human resource management come into practice in direct or indirect factual IP and know-how protection. Reflecting the findings on factual protection with the human resources management research, three factual protection levels in the context of human resources management can be identified: a *human-control, human-motivational*, and *human-educational* level.

4.3.1 Human-Controlling Protection Means

Controlling protection means comprising a *physical* and a *logical* control. The former secures the access to the firm's facilities; the latter secures the access to information. Both controls are realized by means of a role-based structure, which contains different security levels for different job descriptions determining their physical and logical access rights.

Physical Access Control Physical access control refers to protecting substantial corporate facilities and values. It represents an integral part of building security technology. A set of corporate regulations defines which person is authorized to enter secured areas such as facilities, buildings and rooms at a specific period of time. Employees are equipped with means of identification such as token-based means or use biometric attributes for verification to get access to a secured area.

Token-based means of identification include contact-less or contact-based plastic cards so-called smart cards, key fobs or plastic bracelets. They typically comprise an electronic, tamper-proof chip, which holds an identification number or a specific access authorization. At an access gate, a corresponding verification device (e.g., a contact-less reader) is validating the token.

Biometric attributes such as fingerprints, an iris, or the hand's veins can serve as means of identification as well, either substituting or supporting token-based identification. A corresponding verification device attached directly at the access gate validates the biometric attribute. Generally, the biometric access control is used for high security levels only.

A physical access system follows a specific organization, providing a logical structure according to the operator's requirements. In order to optimize the level of security, physical access control installations follow the principle of concentric circles with increasing security mechanisms. The outer-inner-circle-principle determines the access rights depending on the security demands of different locations. An outer circle consisting of a low security level such as the company campus and an inner circle, e.g., the companies' R&D facilities with a high security level usually determine the access requirements. Typically, a role-based structure determines the physical access control. A role provides a set of authorizations to enter specific areas, and that role can be transferred to a specific person within the system. Corporate premises usually consist of areas with differing requirements to physical security (e.g., reception, canteen, meeting rooms, office rooms, IT rooms, R&D offices and labs, safe rooms ...). Next to gates, cards, and verification devices, a physical access system typically consists of a background system that is connected to all verification devices through a network. Security personnel operate the background system by storing and managing the access rights of all the people using the system.

Logical Access Control Logical access control encompasses the protection of corporate values stored in computer systems. It refers to software safeguards, ensuring that only an authorized person can access specific information and data. It is an integral part of computer security. A set of regulations defines which person is authorized to access specific information or perform specific actions on a computer system. A person is represented by user identification, which is a unique identifier on a computer program or system. In an authentication process, the user is verified by the computer system or program attempting to confirm his or her identity. The user identification necessary for the authentication process is represented by a token, e.g., a smart card or USB token or simply by a user name. Both generally require the person to provide a password or PIN. Logical security is typically used in different applications to secure data and information: (1) Computer log-on; i.e., getting access to specific computer program or system; (2) Message authentication and encryption; i.e., a message is being received unaltered from the correct sender and unreadable to anyone else than the correct recipient; (3) Single-Sign-on; i.e., secure log-on to a set of computer programs and systems with a single token or user ID and password.

Logical Outflow Control Controlling the outlet of information is challenging and cannot assure a complete protection. However, barriers for fraudulent outflow of information can be increased. Logical outlet control comprises the outlet due to fraudulent behavior of employees as well as outlets due to lost or stolen computers or information storage devices. For the latter, basic solutions are secured USB memory tokens, safeguarding IT solutions such as digital rights management applications or self-destroying applications in case of wrong password entry. State-of-the-art encryption mechanisms in firms include Advanced Encryption Standard (AES) ciphering with key lengths up to 256 bit. A typical application is Adobe's Acrobat software that provides users with printing forbearance mechanisms and document password authorization for individual documents. Corporate travel regulations should also consider the forbearance of information transportation, e.g., either no or limited storage of data on a laptop or USB memory token, travel laptops without any information or no laptop and storage transportation at all. The internal outflow of information however is very difficult to detect and to proof. Thus, to limit the outlet of information the intrinsic motivation of employees plays a key role, which complements the human controlling means.

4.3.2 Human-Motivational

In a round table discussion with the Chinese government, 59 % of the participants from foreign invested companies stated that their most significant problem is the recruiting and retaining of managers (Bjoerkman and Lu 1997). Many problems concerning know-how drainage and the loss or misuse of IP can be associated with problems in the area of human resources management. Particularly performance issues and staff retention are challenges that mainly address employees' motivation (Child 1994; Henley and Nyaw 1990; Nevis 1983).

Western concepts of human resources management and in particular the motivation of employees can only be transferred partially to a country where employees have been motivated differently—now and again only to do what is best for their country. However, over the past decades Chinese enterprises have gone through a number of market related reforms in order to create and follow profit and productivity objectives rather than ideological, political or specifically social goals (Boisot and Child 1988; Chen 1995; Jackson 1998; Walder 1986). China's changing social and economic infrastructure is accompanied by changing motivational motives. Retaining key employees is crucial to reduce the drainage of know-how and thus to factually protect IP. The elaboration of motivational techniques in the Chinese context needs to be considered and has been developed and employed. The way Chinese companies motivate their employees can be understood within Katz and Kahn's motivational patterns of rule enforcement, external rewards, and internalized motivation (Child 1994; Tung 1991) and has been employed in the Chinese context by Jackson (1998). Rule enforcement addresses the fact that employees obey rules because they have been set from a legitimate source of authority and can be enforced. Thus, the possibility to enforce legal contractual agreements influences the motivation of an employee. Complementary to contractual terms and conditions and non-disclosure agreements, an IP policy as well as the signing of a non-disclosure understanding in a personal briefing before leaving the company should be common practice.

The motivation of employees also depends on the work value system, which varies between western countries and China. Cyr and Forst (1991) argue that Chinese workers are shifting towards a value system that is more goal-oriented. Many interviewees reinforced the evidence that money is an important part of a reward system. Successful headhunting and poaching of one's key employees is often based on higher wages. So a main challenge is the employee retention. Thus, monetary incentives are essential for the recruiting and retaining of key people. In China other important aspects to attract and retain key people are important hygiene factors such as welfare packages (Jackson 1998).

Several interviewees stated that they support their key employees with rent allowances, provide housing or even support vocational training of key employee's children, e.g., with inexpensive or tax-free loans for (international) education. A multi-national company reported to provide a 7-year tax-free loan for personnel on the top two management levels provided that the employees stay for at least the same period of time within the company. Another company attracts a key employee with the provision of a large apartment in Shanghai, which had been anticipated as important not only for the employee but for his wife and child and its educational opportunities. Thus, on the one hand such measures serve as motivator to start working at the company, on the other hand this is one mean to retain key employees and to increase the reputation of the firm mid- to long-term. Such welfare packages are important HR instruments in China, since the provision of housing is common practice in state-owned companies. Since Chinese employees would give up benefits when moving to a foreign company (Jackson 1998), such measures become relevant to win and keep the best talent. For young Chinese employees it may not seem to be so important since they often live with their parents. At the same time it puts pressure on companies, which do not provide such housing opportunities or associated benefits and has implications for recruiting experienced key employees.

Next to welfare packages creating opportunities for employees to climb the corporate letter generates internalized motivation. Another important aspect are prestige objects that represent their status within the firm. Chinese employees are used to lower standards from state owned companies; however, their expectations of the working environment are generally high if working for a European company. In this vein, one company uses the provision of a valuable company car to feed the perception by employees of cars as a status symbol. Other external rewards are also company computers and smart phones, which have been mentioned as important status symbols for employees. Non-materialistic career opportunities are of a substantial character, too. In western companies the prospect of an international career path compared to local competitors is better. Compared to Chinese SME this is one motivator, which promotes western companies with clear options for an international career. The corporate identity and the corporate social responsibility are further hygiene factors that attract new employees. The visibility and social engagement of a western firm is important, too.

Example:
IP guideline for a transparent reward and remuneration system in China

Rewards and remuneration for inventions

1. The company will pay its employees as a reward and remuneration for an invention he/she created:
    a) A one-time payment of RMB 5'000 Yuan per patent application at the time when the company submits a patent application related to an invention to the Chinese Patent Office.
    b) A one-time payment of RMB 5'000 Yuan per patent at the time when the Chinese Patent Office grants a patent for an invention.
    c) One-time payments of RMB 10'000 Yuan per patent five (5) years, ten (10) years, and fifteen (15) years, respectively, after filing of said patent with the Chinese Patent Office, if said patent is still in force at that time and the company still uses the invention covered by said patent (directly or by licensing). If a patent is not yet granted the stated years after filing, but is granted later, the company might postpone the payment until the patent is granted.

2. If a patent names more than one inventor, the company shall allocate the payment to the inventors according to their shares as fixed in the inventors' declaration. The inventors shall jointly sign the declaration and shall clearly state the share (in percentage) of contribution by each individual inventor. If no such declaration is provided, the company is not obliged to make any payments.

3. The rewards and remuneration shall apply to inventions which the employee makes
    a) in the course of performing his/her duties; or
    b) in execution of any task, other than his/her duties, which the company assigned to him/her; or
    c) by using the material and technological means, or the experience of the company, or
    d) within one year from the termination of his/her employment with the company, where the invention relates to the employee's working duties or other tasks assigned by the company during the employment.

4. The company shall effect the payment together with the annual bonus payment for the same calendar year.

5. In case after his/her employment with the company the employee does not notify the company of a change of his/her address and bank account, the company shall not be obliged to make any remuneration payment under this article.[x]

Corporate social responsibility contributes to the identification and pride within the company and engenders commitment, greater morale, dedication to excellence in performing work tasks, job satisfaction, as well as feelings of well-being (Tymon et al. 2010).

The internalized motivation is influenced by reward systems and the reputation of the company. An IP policy allows a transparent and common practice of IP-relevant issues. It determines the remunerations, which are fulfilled as a one-time payment for a patent application and a patent grant. In the following, an extract of such an IP policy established by a European multinational company is presented.

Next to the remuneration, an IP policy can give the necessary transparency about regulations after the end of an employment. The presented IP policy is designed for all R&D employees in China within that company.

Compared to western management styles direct criticism should be avoided. Instead, the message should be carried in a kind wording to prevent the loss of face. A manager reported that the guanxi to his employees is important and requires more efforts compared to his western work experience. On a birthday he goes for dinner with the employee or calls him at home. Managers build up trusting relationships, which compared to western teams need more personal efforts for a sustainable internal guanxi.

The principal of internal guanxi can be compared to western relationships such as family, friends, colleagues or other social networks. It is a similar concept, however, the difference is that in China the network is considered truly important to help in business matters and is often more intense compared to western relationships. Internal guanxi is an open system of connections within the firm, which means that the connection to a friend of a friend is provided and seen as the opportunity to improve the guanxi with both involved parties. The open system represents a challenge when it comes to the protection of know-how. The guanxi of employees beyond the own company relies on favors that are given in a bi-directional way, thus the risk of know-how drainage is crucial. A continuous training of IP, a continuous communication of its importance as well as the underlying legal contracts and obligations are essential. Reminders about particular issues must be repeated systematically, or Chinese employees will think that the issue is not relevant anymore and will ignore it (Keupp et al. 2010). Such trainings shall also put forth why the protection of IP is important, to understand that if IP is given to third parties, it will hurt the company in the first place, but it also has a retroaction for the employees.

Once established and maintained, the internal guanxi can also be exploited. The employees' guanxi bears business opportunities, which stay or leave with their retention. Paradoxically, the company's own internal guanxi can be used to exert pressure. In one case a manager identified that his employee had disclosed key know-how and confidential information to competitors without authorization. The manager forced that employee to talk to the competitors in the company's name and issue a threat that the firm would retaliate. As reported by the manager, the person became isolated by his own internal Chinese network, and was treated as a traitor, which can lead to a breakdown of the personal network. This prospect is fearful for most Chinese, whose society is based on networks of personal relationships that are paramount in personal career paths (Luo 1997, 2007; Park and Luo 2001). Consequently, the other employees of the firm know the consequences in case of IPR disclosure, which makes them very likely to abstain from such behavior. The first facet of internal guanxi, the trust building and relationship network confirms that a foreign company's level of trust towards its host-country employees relates positively to those employees' loyalty (Child and Möllering 2003). If employees establish a positive connotation of the company due to the described measures (rule enforcement, external rewards and internalized motivation) they are more likely to display loyal behavior. Thus, managers who employ the human-intrinsic measures make use of China's dependence on social relationships and cultural values and norms to protect the company's IPR. In contrast, the introduced facet of exploiting internal guanxi relies on mistrust, which makes it a 'classic example' of a principal-agent relationship (Keupp et al. 2010). Although personally threatening employees with the loss of their personal network may be questionable from a business ethics perspective, it is, according to the interviewed managers, an effective prevention of the outflow of know-how.

4.3.3   Protection by Human-Educational Means

Since Chinese employees tend to require a stronger need of security and clear instructions for individual tasks, such rules need careful implementation. Otherwise they might become more important than the goals they were designed to accomplish, causing employees to follow the rules for the sake of the rules themselves (Hall and Ames 1995; Jackson 1998). Most likely employees in such a system become increasingly unable to operate on their own initiative. Companies report a reluctance of local employees to seek responsibility or to show initiative and to participate in decision-making. They also report a need to closely supervise staff and to train the ability to be creative and to proactively explore technology development and IP awareness.

It is crucial to find a balance between defined rules and instructions as guidance and a good confidence level to achieve individual goals and personal success instead of strongly enforced regulations as existent in state-owned companies. The involvement in decision making processes also serves as a motivational measure, however, should be implemented for selected employees only, since the fear of false decision making is deeply rooted within Chinese working behavior but decreases with work experience. Such learned helplessness arises from a prior experience of incapability to be able to achieve future success (Schermerhorn and Nyaw 1990) and relates to cultural behaviors such as uncertainty avoidance and face saving.

Siemens' Chinese CEO stated that a major difference between western and Chinese graduates is their demand of seeking self-responsibility and being part of decision making processes. Although depending on the individual person, Chinese people rather build up responsibility and seek decision taking over time and work experience. In managerial functions age is a far more important indicator for experience and authority compared to western indicators such as competence and performance. Multi-national firms such as Siemens, ABB or Philips foster initiatives and tap their self-responsibility. Such initiatives are often realized within a company on a project level and by IP trainings. The different backgrounds of the attending audience should be considered to address all levels of knowledge. Also the integration of IP issues within an R&D stage-gate-process increases the awareness of IP generation and protection. From a managerial perspective, the education of employees enhances their capabilities and competencies. Financing trainings and other educational subventions of an employee or his/her family education can be used to lock-in the employee. Such a lock-in effect can be realized for a certain period in time. However, it is limited due to the compensation by a subsequent employer.

Multinational companies go beyond the internal education and support public IP education. Philips started to offer courses at several Chinese universities for law and science students. The Philips IP Academy in China was established to support the Chinese government in enhancing the IP knowledge education. Philips has started to set up three local IP Academy projects since 2004. The aim of the IP Academy is to share international IP expertise with Chinese students, increasing IP knowledge and awareness in China. The Academy includes IP courses given by IP experts from Philips, the EU and the US academics. A scholarship program, an IPR research program and an exchange program for EU and Chinese professors are an integral part of the program. The three universities involved are Renmin and Tshinghua Universities (both in Beijing) and Fudan University (Shanghai). About 650 law and science students from these three universities follow the Philips IPR courses (Philips Corporation 2009). Topics that have been taught so far are IP law, patent law, copyright, trademarks, design rights, trade secrets, international IPR treaties and the role of patent attorneys. Philips has developed specific course material for these courses such as a syllabus, presentations and law books. Further modules will be added to the program. The objective of such initiatives is to broaden the knowledge about IP and its importance at an early stage. The development of local IP competencies is essential since the war for talents already exists for IP experts in the Chinese market. There are still too few employees who bring a certain amount of IP education and IP work experience with them.

4.3.4   Summary Human-driven Protection

The identified human-driven protection means directly or indirectly reduce the risk of know-how drainage. Controlling the access to facilities and to data is a basic barrier to loss of crucial data. However, the drainage of information is difficult to control, thus, employees should be intrinsically motivated not to unseal confidential data. Using Katz and Kahn's motivational patterns, the human intrinsic protection means can be described as rule enforcement, external rewards and internalized motivation. Due to cultural differences the means for intrinsic motivation differ from western practices. Welfare packages are only one example of a common incentive at state-owned firms, and needs consideration in western firms to win and retain key employees (Fig. 4.5).

Furthermore, the internal education of IP matters, their transparency and importance is the baseline for IP protection. Educational initiatives and career prospects can be established to increase the reputation of the firm to become an attractive employer that wins and retains the best talent, and that includes IP experts. The concept of internal guanxi helps to increase loyalty and commitment to the firm and should also be exploited to deter IP theft.

Case Example: Counterfeit Drugs

In many cases counterfeit products are of a lesser quality. This can be annoying—for instance when a pair of jeans wears out quickly—when it comes to drugs however, it is often hazardous to life. Counterfeit drugs can look exactly like the original product but frequently contain far different ingredients. It is not unlikely that a nutritional pill contains a toxic dose of chemicals. The WHO groups counterfeit products into six categories (WHO, 2011):

1. 32.1 %: Products without active ingredients
2. 20.2 %: Products with incorrect quantities of active ingredients,
3. 21.4 %: Products with wrong ingredients,
4. 15.6 %: Correct quantities of active ingredients but fake packaging
5. 1 %: Copies of an original product
6. 8.5 %: Products with high levels of impurities and contaminants

Many counterfeit drugs are sold online, claiming to be the original product. Customers have a hard time distinguishing them. High drug prices make the online offers seem appealing. The Swiss campaign STOP PIRACY tries to raise awareness for this dangerous topic. In late 2011 the campaign placed billboards all over Switzerland and offered tests to verify the authenticity of drugs bought online. The following picture shows a Chinese production facility of counterfeit drugs. Drugs produced in facilities like this can be bought all over the world.

© Pfizer | www.stop-piracy.ch

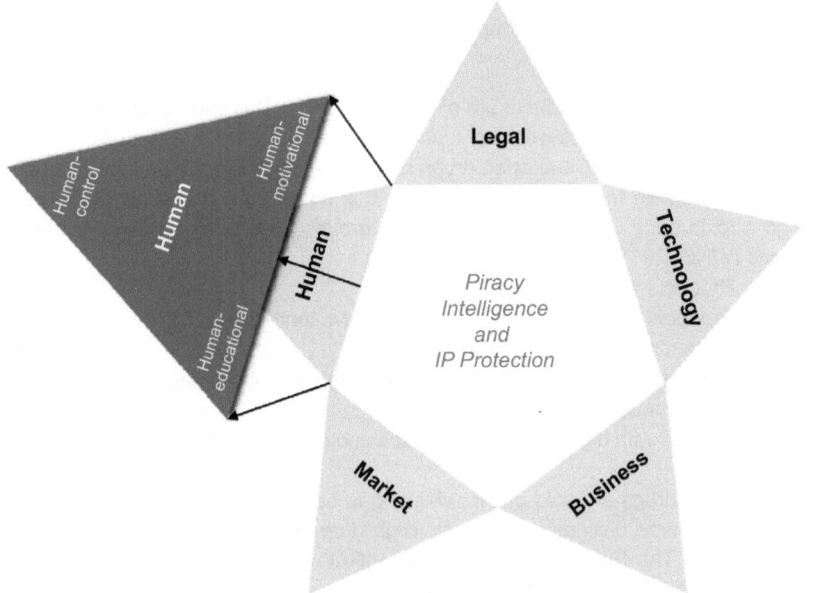

Human-control
- Physical access
  to facilities
- Logical access control to data
  for internals
- Logical outflow control

Human-motivational
- Rule enforcement (IP policy,
  clear expectations)
- Rewards (inventions, patents,
  project success)
- Internalized motivation
  (career opportunities, welfare
  packages, *internal guanxi*)

Human-educational
- Internal IP and
  product education
- External IP and
  product education

Fig. 4.5 Human-driven protection means

4.4   Technology-driven Protection

One mean to counteract IP threats is the application of technology-driven protection means. Particularly to reduce the risk of product piracy, protection technologies applicable to a product in a visible or invisible way have gained importance. Also machine-supported technologies such as RFID tags are in use. Besides the applicable features for enhanced protection the intrinsic protection of a technology can be useful and exploited as a protection mechanism.

Technology management in China needs careful consideration which know-how will be transferred to China, how it is shared in cooperation with partners, how suppliers are involved and how it is managed internally. Platform and module management gain importance in emerging markets. Technology-intensive modules have to be designed as black boxes and handled somewhat like national treasures. These technology-driven aspects derive from three dimensions of protection: *a technology-applicable, technology-intrinsic* and a *technology-managerial dimension.*

4.4.1 Application of Protective Technologies

A variety of protective technologies is available; however, these are often not used until a firm faces sales of counterfeit products. The application of protective technologies can be clustered in three different categories (Fuchs 2006):

1. Application of visible protection technologies
2. Application of invisible protection technologies
3. Application of machine supported protection technologies.

Visible protection technologies are applicable directly on the product, spare part or packaging and are of visible nature. Some examples used in practice are:

• Holograms, e.g., embossing holograms, polymer holograms
• Optical Variable Devices (OVD); Diffractive Optically Variable Image Devices (DOVID)
• Foils (e.g., transparent diffractive foils)
• Security labels and security seals (e.g., self destroying seals)
• Security paper, ink and printing (guilloche techniques, UV colors, thermo-reactive colors, color-shifting inks, typographic copy protection)

Invisible protection technologies can only be detected with technical accessories. These technologies are microscopic color particles, special ink reacting to ultra violet light, fluorescent and absorbing color or invisible barcodes. The underlying technologies are often:

• Micro-colors
• DNA and DNA computing
• Nano-technology and nano-biotechnology (e.g., bio-code-labels, bio-id color)
• Isotopes
• Chromogenic systems

Machine supported technologies are relying on data, which have to be distracted with technical receiving and/or sending tools:

• RFID tags
• Barcodes
• Smart cards
• Optical Character Recognition (OCR)
• Biometrics

- Cryptology/code seals (physical or IT supported identification, security codes are managed in a closed system and can be checked online)
- Digital watermarks
- Chemical markers

Due to additional costs, the application of protective technologies is often reduced to a minimum of protection, realized for example with barcodes or holograms. Stihl applies a protective technology-applicable measures. For instance, they have used bio-code labels, which are not visible and do not necessitate the change of the label itself. For testing of originality the label has to be moisturized with a test liquid. The advantage of these labels is the possibility of multiple testings, which allows a continuous testing along the distribution chain. To mark even small components, Stihl uses the trademark protected stylized S embossed on spare parts.

Danfoss has substituted one-dimensional barcodes by two-dimensional barcodes using it to control the logistic chain. Instead of sticking a company label onto the product, the brand name is embossed on it. RFID tags are commonly used for tracking and tracing along the logistic chain or to identify the inventory (e.g., Walmart) with product protection as a side effect. Pharmaceutical companies are securing their packaging and blisters with security seals or labels. A European automotive OEM uses numbered hologram labels. To ensure that the OEMs suppliers do not produce any factory overruns, the OEM requests the application of the numbered holograms on important parts. The holograms are not ordered by the supplier himself but provided by the OEM. The numbered holograms are counted and allocated for each supplier, which allows technical protection of the good and additionally a certain control of the supplier. In combination with the serial number of each part the identification and the tracking of a part manufactured by a supplier is thereby secured.

Due to additional costs, the application of protection technologies is often reduced to a minimum. The consideration of technical features as mentioned above should be considered in a broader way to exploit other opportunities. Such opportunities are manifold, e.g., tracking and tracing for supply chain support, optimization of the logistic chain, support for inventory management, identification of products for maintenance with relevant documentation, health and safety instructions and user manuals. The identification of information that is beneficial for the customer is often neglected for exploitation via barcodes. New applications such as mobile applications for barcode scanning already exist, e.g., in the fast moving consumer goods industry. The general information of a product as well as specific information such as its ingredient or health information or links to other websites are stored on a server. Thus, considering the overall functionality and application potential of protection technologies can foster their application. The integration of verification for originality is an integral part complementary to other customer benefits. For manufacturing firms the integration of further customer benefits, such as product information, maintenance reports, as well as health and safety instructions relevant for inspectors could be amongst others the initial reason to establish smart barcode solutions or RFID applications. The inherent protection opportunities would be of additional benefit.

4.4.2 Technology-Intrinsic Protection Means

Technology-intrinsic protection means describe the protection due to the inherent protection character of a technology or a system. Technology-intrinsic protection can be achieved with a high complexity of the product, its production and/or its service offerings and its maintenance. Another important differentiation and protection potential is the customization of products. Individual solutions increase the imitation barrier. The lock-in effect can also be realized in the manufacturing industry, where system solutions provide a strong commitment of the customer. Other competitors are locked out due to the necessary know-how of a certain service or due to a specialized system.

The degree of complexity of a product or process technology impacts the efforts of imitation. Amongst others, the product and production process complexity is influenced by the degree of standardization, number of interrelated elements, degree and complexity of integration, number of different (proprietary) hardware and software parts, number of processes involved, difficulty of each process, diversity of processes in distributed (international) networks and number of suppliers.

The technology complexity is a key protection mechanism for Instal's industrial installations. The installations are composed of hundreds of modularized components with a high degree of technological complexity and process know-how. The high degree of implicit technological know-how inhibits simple imitation. If a competitor would succeed in copying one component, it would still be difficult to replicate all components needed. Particularly the process design is unique and is based on the engineering expertise of the worldwide engineering network. This is one reason why the company can maintain a lead-time advantage, which is exploited by introducing the newest technological generations in China. The innovation leadership allows a temporary protection and first-mover advantage for the company while competitors with less expertise suffer from a comparative disadvantage.

Builtsys' (an industrial leader in supplies for the machinery industry) technological specialization is realized by adaptation to the local needs such as robustness and durability of their machinery. Due to their quality advantage, their major competitors in China—although many Chinese competitors exist—are European firms. Due to a high amount of technological state-of-the-art in the machinery, specific IP for machinery features allows the company to significantly upgrade their products. The competition in China derives from European firms, which also have a strong IP portfolio.

For a wide range of complex, high-technology goods, such as chemicals, electronics and machinery, the costs of imitation average 65 % of the costs of innovation. Thus, the costs of complexity can be interpreted as a tax on imitation (Glass and Saggi 2002). The technology-intrinsic complexity and the technological customization raise the imitator's costs, so that imitation even if technically possible becomes on the same quality level economically unviable. The knowledge needed, complexity and ambiguity of resources create barriers that inhibit imitation (Reed and DeFillippi 1990; McGaughey et al. 2000).

4.4.3   Managing Technology and Technological Know-How Distribution

Modular products distributed across different development and manufacturing facilities reduce the risk of know-how drainage. A black box principle protects modular products by the production of technology-intensive components in countries with a strong appropriability regime.

The assessment of technical know-how is crucial for technology management in China and the transfer of critical know-how needs careful evaluation. If the risk of know-how loss and, as a consequence, the risk of losing the competitive edge is high, it can be advantageous not to transfer the know-how. Before entering cooperation, the know-how evaluation helps R&D managers to determine how to share which kind of information. A setting with multiple suppliers allows the distribution of know-how and the reduction of dependency on a single source. The 'need-to-know-principle' with restrictions on knowledge and documentation transferred to suppliers ensures that not more information than necessary is available to third parties. In contrast to the restricted know-how transfer to China, recently the problem of getting know-how out of China for international R&D subsidiaries has occurred. Strict regulations in China demand to keep know-how within the country. Inventions made in China have to be patented in China first. Furthermore, the protection of the production process via secrecy is one mean to reduce the risk of knowledge drainage. However, the patenting of production processes offers a legal basis for enforcement in case of the fraudulent process exploitation by a third party. Including production details, which are not relevant for the technical differentiating feature can be used to identify counterfeit parts and to proof the identical copy due to patent documentation (Fig. 4.6).

The Weidmüller Company—a manufacturer of electrical connectors and transmitters as well as signal and data electronics—was able to prove an identical copy of their product due to the integration of small extensions that are necessary for their internal material flow in the production line. Since the Chinese copy was obviously manufactured in a different way but still included the same extensions, it was clear, that the patent document was used it and it thus infringed the patent protected product. Thus, technical features integrated in the patent application to confuse imitators can be used to proof the identical copying of a component. This may also strengthen the legal options for successful enforcement of own IPRs.

4.4.4   Summary Technology-driven Protection

Technology protection measures are more than protection technologies in a visible, invisible or machine supported way. The intrinsic protection due to the complexity of a technology or process, the customization of products as well as a lock-in effect by system solutions increase the barrier of imitation. Managing the crucial technical know-how includes decisions on transferring it to China or retaining it in stronger appropriability regimes.

The black box principle is a solution, which combines both transfer and non-transfer of production and technology to China. Depending on the degree of tacit

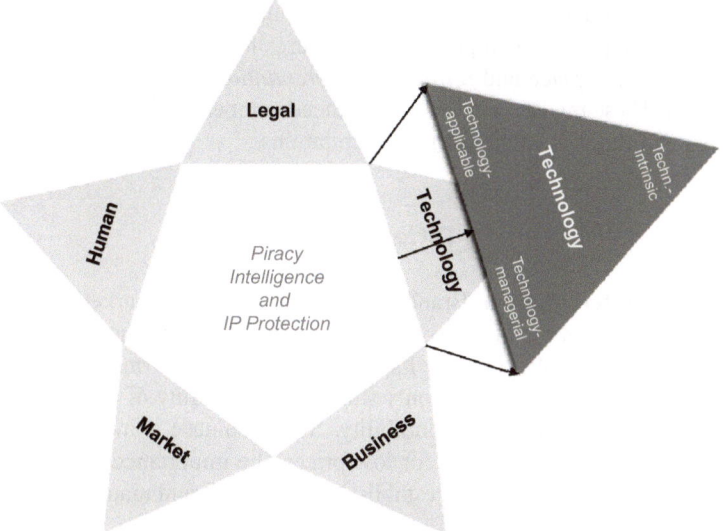

Fig. 4.6 Technology-driven protection means

Technology-applicable
- Visible
- Invisible
- Machine-supported

Technology-intrinsic
- Complexity
- Customization
- Lock-in effect

Technology-managerial
- Know-how distribution
- No transfer of crucial know-how
- Black box principle
- Distribution among different suppliers
- Lead-time advantage through process know-how advantages and newest technology transfer
- Know-how evaluation and need-to-know principle

know-how and realized lead-time advantages due to enhanced process know-how, the transfer and sales of newest technologies offer a competitive edge. The evaluation of know-how and the consequences in case of loss are essential for such decisions. Before entering cooperation with partners the consequences should be assessed and a clear decision has to determine which know-how will be shared and distributed across partners.

4.5   Business-related Protection

The risk of losing know-how in China challenges the decision whether or not to transfer knowledge and technologies to China at all. A clear preventive strategy is not to transfer key know-how. However, often the competitive situation requires such a transfer and the protection can only be realized by establishing competitive advantages, which protect the business. Competitive advantages can be unique business models or strategies of a firm, which inherently realize a protective function. Examples of such protection measures include *brand and service differentiation, quality, price* and *reliability differentiation* as well as *controlling the supply chain.* These measures allow an enhanced competitive advantage towards competitors and can protect the firm from imitations.

4.5.1   Protection by Differentiation

Brand Differentiation Brand imitation is a valuable marketing strategy based on the exploitation of similarity of a genuine product with the intention of facilitating the acceptance of a brand. Especially for western firms with established brand management, the brand not only stands for the quality of the product but it signals competence, prestige and durability. Since Chinese competitors are catching up with good quality products for low prices, the importance of brand differentiation increases. Companies have established different brand management measures:

Service Differentiation and Service Branding In the manufacturing industry companies position themselves more and more as service providers instead of pure product providers. Particularly consultancy services and customized products are individualized offerings that are difficult to imitate. In addition to well-known products, complementary services are offering differentiation potential. The branding of these services becomes as crucial as the branding of products. Once a guarantee and service are the selling point for the customer, the service has a profitable benefit for the company and customers of fake products are automatically excluded from such offerings. As standard services remain free of charge or at a low cost price, premium offerings with certificated or quality controlled services are complementary modules. The machine tool manufacturer Trumpf offers its customers in China a job-shop where they can try the machinery before purchase. The job-shop allows a real-time production of parts and the training to operate the machines. This also allows a demonstration of the underlying quality of the resulting end product.

Strong Corporate Branding Especially in the industrial goods industry, the corporate brand can replace a product brand. Thinking of a product, the customers' connotation relates to the corporate brand instead of the product brand. Experience shows that Chinese customers are buying a 'Siemens' or a 'Stihl'. The strong corporate brands enhance the service branding which is often a combination of the corporate brand name and a service indication.

Ingredient Branding Companies enhance the status of their product with China relevant ingredients. An ingredient is a component of the product, which has its own brand identity. Similar to a co-branding with well-established firms, the ingredient branding upgrades a product by the usage of a positive connotation of another brand. The most common ingredient branding is the example of 'Intel inside'. In the industrial goods industry the use of ingredient branding can be exploited, too. The indication of certificated independent services, quality or safety approvals can be used for upgrading the own product. Also the usage of marketing the associated number of patents inherent in the product, as an ingredient is one means to underline the quality of the product. Depending on the value proposition of the product, ingredients are often used for the differentiation to competitive products.

Quality, Price and Reliability Differentiation The differentiation in price is one indicator for the customer whether or not the product is fake or original. The prerequisite is the deliberate knowledge about the quality of the genuine in combination with its price. Assessing the price and the quality of an imitator's product compared to the original product, four different characteristics can be determined:

1. Low pricing and low quality
2. High pricing and low quality
3. Low pricing and high quality
4. High pricing and high quality

The mediating factor for these characteristics is the degree of deceit. In the case that a customer purchases a product of high quality with a high quality of imitation, he believes to have purchased an original. A typical example of such sales are overruns, which are products that have been fraudulently produced without the genuine manufacturer's permission, e.g., in a third shift of a supplier. Due to the high quality of the imitated product, the original manufacturer can suffer significant loss in sales if the product is sold with a high price tag. Since the imitated product offers a similar differentiation compared to the original, the customer might be willing to deliberately buy the imitated product, which will also impact the sales share of the original producer. In the case that the quality is low, the imitation might not attract the customer base but can significantly cause legal issues. That is the case, if a customer tries to make use of the guarantee or in case of claiming health issue damages. Albeit not responsible, the original manufacturer suffers from the illicit sales and can be confronted with compensations. In case the pricing of the imitated product is low, the quality will determine the impact for the original manufacturer. In case of a high quality, the differentiation of the original offering, e.g., upgraded by service offerings or spare part guarantee, will determine the unique selling point. In case of a low quality the impact can be the dilution of the brand. Since imitated products congest the market, the exclusivity of the genuine brand is diluted. The latter case has been experienced in China within the consumer goods industry as well as within the fashion trade (Fuchs 2006).

A way to counteract such imitations is the quality differentiation by technical features. Performance, durability, ecology, services, system selling, technical consulting and expertise amongst others are common differentiation parameters. Thus, the education of the customer about the unique selling point is essential. The product portfolio and its fit to the Chinese market, customer needs, and the competitive situation should determine the pricing strategies for the Chinese market.

Reliability Differentiation Similar to the quality assurance, the reliability of the product and its quality creates a differentiation potential. Particularly, in the case of long product lifecycles it is essential for the customer to have the assurance that the manufacturer will exist in the future. The availability of maintenance services, technical upgrades and spare parts are relevant features during the product's lifecycle. Also the assurance of possible warranty claims can significantly influence the purchasing decision of the customer. The reliability differentiation is difficult to copy due to its intangible and long-term character.

Differentiation by Innovation Speed and Product Lifecycle Management The lead-time advantage renders the innovator a temporary period of time for the market penetration. The product development time determines the lead-time advantage. The imitator's development time determines the market entrance time of its imitations. Anticipating that the development time of the imitator is shorter than the regular development time of competitors, the imitator will launch the products in a shorter time than third parties. Thus, the imitator becomes a fast second mover (with an advantage with respect to competition). To slow down imitation, the degree of intangible assets such as key knowledge, e.g., about production processes, installation concepts, technical consultancy inherent in the product or system offering has to be high. In particular, the conceptualization and the configuration and adjustment of process chains are based on tacit knowledge, which is difficult to copy.

Strategic Release Planning A simultaneous product release in all markets can encourage a lead-time advantage since regionally different launches may allow potential second movers to reduce the lead-time for other markets. Another strategy is a release planning which allows an innovator to launch his products in a timely manner to stay ahead second movers. Incremental innovations are launched in shorter periods of time. A European cosmetic company established a strategic release planning, which allowed the introduction according to the occurrence of imitations. The ready-to-launch strategy facilitates the reaction with a new product introduction as soon as imitations occur on the market. The objective is that imitations always copy former product generations and the innovation leader maintains its market share.

The transformation of that strategy to the technology intensive industries is limited due to long development times and product lifecycles. However, this strategy can be applied for spare parts, low-tech products or the strategic release of marketing or service activities. Complementary consumables sold in system solutions could be considered for that strategy as well. This strategy requires a thorough market monitoring that ensures an early identification of copies in the market. Thus, the ease of identification of imitations is crucial to clearly separate original from fake, e.g., by means of additional technical protection means.

4.5.2 Protection by Operational Control

The control of the own processes is essential for reducing the risk of know-how drainage and fraudulent exploitation of proprietary assets. Typically, the supply chain bears the risk of IP threats. The own sales processes can be a weakness or a strength when it comes to protection against IP theft and imitations.

Lot-Size Control The determination of certain lot sizes in the production as well as the suppliers' sites enhances the controllability against overproduction. With technical protection means such as holograms, each production lot has its identification and can be tracked. Daimler purchases holograms and provides its Chinese suppliers with such labels. Daimler applies that strategy to be able to track their products but also to assure the right quantity within the production. Another advantage is the proof of originality in the case of safety-relevant parts. This strategy is often combined with the quality assurance measures. Regular audits of the supplier support the outcome of the right quality and quantity of an order. However, audits shall not only fulfill a control function but a continuous personal contact with the supplier. An outcome could also be the necessity of technical support to realize the required quality, which enhanced the controlling of the supplier and a reduced risk overruns.

Controlling the Supply Chain Imitations often occur at some point in the supply chain. Tracking and tracing of products, e.g., with protection technologies such as RFID, facilitate the identification and tracking of products within the supply chain. The export of imitations are challenging for genuine manufacturers. Cooperation and trainings with Chinese customs are essential to identify and stop such fraudulent exports. Many firms closely collaborate with the European customs to provide training and education about incoming goods, which are imitations of their products. The identification of imitations in other countries and the retracing of their origin provide insight into production.

Direct Sales Force A direct sales force creates closeness to customers and can enable the protection against imitations. On the one hand the direct sales force carries out a great part of customer education about the benefits of the original product as well as the possibility of how to identify an original. On the other hand the closeness to customers allows an indirect upgrade of products due to availability and service offerings. Also, the identification of infringements on the market and the competitive situation can be closely monitored. The indirect protection due to a direct sales force is limited to the own employee's integrity. A strong corporate culture and incentives not to act corruptly are the basis for reducing the risk of corrupt employees.

4.5.3   Protection by Prevention: The Avoidance Tactic

A defensive strategy is the avoidance of knowledge and technology transfer to China at all. This strategy needs careful consideration in several directions. The risk of losing market share if not doing business in China, as well as the thread of imitation exports, oppose the risk of losing key know-how and technology lead-

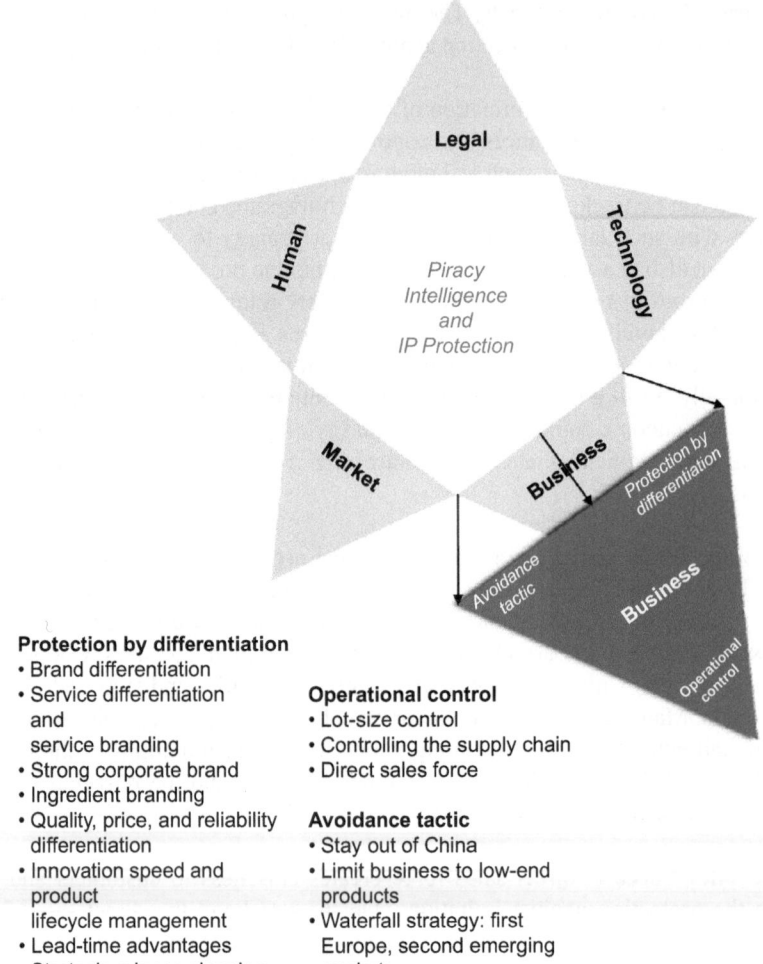

Fig. 4.7 Business-driven protection means

Protection by differentiation
• Brand differentiation
• Service differentiation
  and
  service branding
• Strong corporate brand
• Ingredient branding
• Quality, price, and reliability
  differentiation
• Innovation speed and
  product
  lifecycle management
• Lead-time advantages
• Strategic release planning

Operational control
• Lot-size control
• Controlling the supply chain
• Direct sales force

Avoidance tactic
• Stay out of China
• Limit business to low-end
  products
• Waterfall strategy: first
  Europe, second emerging
  markets

ership when transferring key technologies to China. A differentiated perspective concerning the business success versus the business risks has to be addressed on a case-by-case basis. As such, the entire Chinese business can be avoided; however, the risk of Chinese imitations remains. Yet, only the manufacturing or certain product lines can be transferred to China. However, experience shows that customers in Asia demand the newest product generations. In case the business success exceeds the burdens of IP infringements or imitations the transfer of technology is justified. Siemens China's CTO refers to a win-win situation, which is the overall objective of cooperation in China that can recoup infringements. While Microsoft suffered from the Chinese piracy, a bandwagon effect facilitates the spreading of their software. The company formerly claimed a win-win situation. However, in 2010 Microsoft's CEO reviewed the market development and with less revenues in China then in India, Indonesia and Thai-land, Microsoft will focus on other Asian markets in the future. IP protection is emphasized stronger by the government in these countries. Depending on the individual case and the circumstances, e.g., the governmental and political relevance of technology for China as well as the location of key accounts and thus the demand of local support, determine a company's strategy. In some cases the avoidance can be the most promising tactic (Fig. 4.7).

4.5.4 Summary Business-related Protection

IP management has to consider a broader scope for IP protection—beyond the direct applications of legal and technical means. The unique value proposition of a firm's product and service portfolio can inherently protect innovations. High differentiation to competitors raises the barrier for imitators. Brand differentiation is often built up over several years. Differentiation by service requires a thorough understanding of customer needs. Setting up a service organization in China is far from easy and requires local skills and investments in logistic infrastructure. Professional human resource management must also help to support the selection of the most suitable candidates. Strong corporate branding not only attracts customers but also talented employees.

Innovation leadership by means of quality and reliability differentiation legitimates price premiums that can then be exploited as protection mechanisms. The right innovation speed ensures the pace making of differentiation, particularly when the newest technologies are demanded in China. Product lifecycle management and strategic release planning are means by which to outdistance imitators and competition. Furthermore, the controlling of business processes, in particular in cooperation with third parties, is essential to deter and identify possible leakages and IP threats. Close collaboration and a direct sales force help to educate the customer and facilitate customer loyalty. If the IP threats and factual challenges outweigh the business opportunities, an avoidance tactic for those parts of the business concerned is recommended.

Integrated Management of Piracy Intelligence

5

How do you implement the IP protection star into the daily business of a company? In many firms, the IP management is an isolated unit with little contact with the daily business. Yet, the management of IP protection, the identification of piracy and the application of countermeasures require the involvement of many roles within the organization. Since IP departments are often organized as central units in a firm and are responsible for all IP management tasks and related services, the requirements of an integrated protection system are not met. In most organizations, an integrated management approach that involves different departments to identify and counteract IP threats is missing. The need for such an approach is often only realized after the occurrence of an infringement or serious IP threat.

A shift from an organizational structure towards a process-oriented management against counterfeiting is recommended in China due to the high dynamic of IP threats and the required flexibility (Fuchs 2006). In conjunction with the IP department, the top management should support the awareness and existence of strategic protection strategies and involve themselves in serious infringement cases. A clear sense of responsibility at different levels must be established. If the employees within the organization are not made responsible for IP protection, there will not be the necessary support for a continuous improvement process. Only clear ownership can create change for an effective piracy intelligence and IP protection process. Continuous improvement includes the idea of giving people clear task ownership for piracy intelligence and the application of IP protection means.

Our integrated management model for piracy intelligence and IP protection is following a five-step approach. In the *first step*, an evaluation of key technologies and know-how is essential, and is *followed* by the formation of preventive protection. In a *third step*, piracy intelligence and IP monitoring activities ensure the early identification of competitors and IP threats. In the case of an IP threat, an evaluation phase (*step four*) determines the perceived risks and helps to determine the factual and legal protection means that need to be applied (*step five*). The findings from the identification of IP threats and from the application of legal and factual protection means help to sharpen the evaluation of further technologies and know-how, which is considered for a transfer to China. For such a continuous improvement process,

O. Gassmann et al., *Profiting from Innovation in China*,
DOI 10.1007/978-3-642-30592-4_5, © Springer-Verlag Berlin Heidelberg 2012

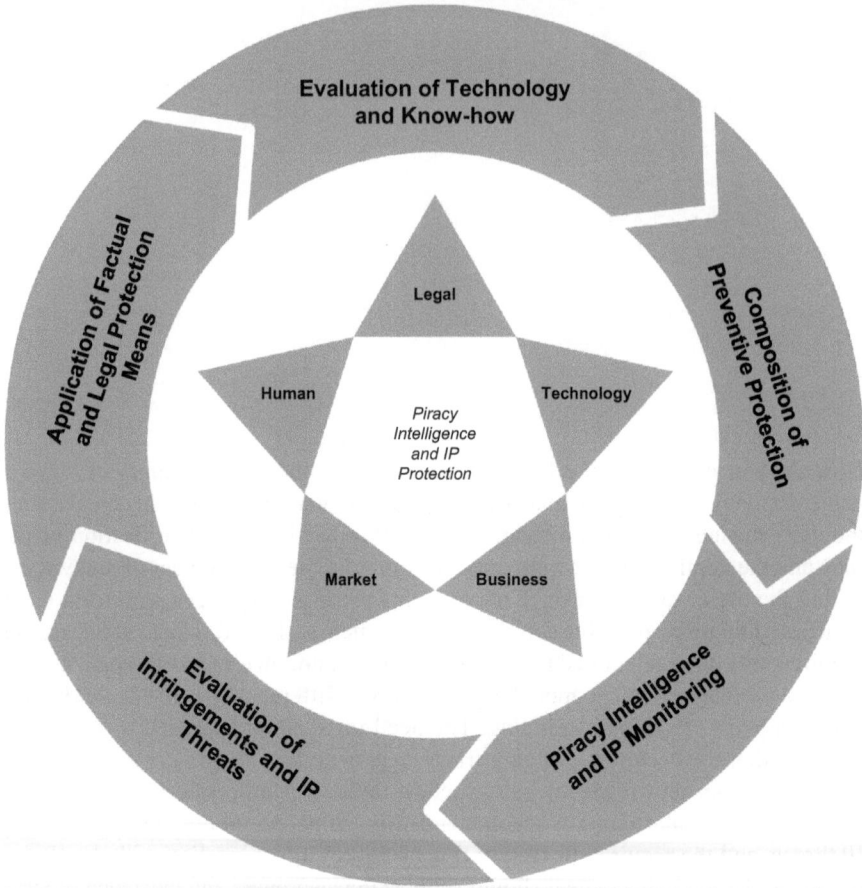

Fig. 5.1 Integrated management model of piracy intelligence and IP protection

the definition of clear roles and responsibilities is necessary to protect effectively intellectual property. Ownership creates change and allows the implementation of a continuous improvement process (Fig. 5.1).

5.1   Evaluation of Key Technologies and Know-How

The baseline for protecting IP in China is centered on the decision as to which technologies and thus knowledge to transfer to China. Firms that have never suffered from IP threats often regard IP threats in China as a product piracy problem that occurs mainly to consumer goods and not their industry. It is essential to generate awareness of both legal and factual IP threats and their impact on business. A situation analysis elucidates what kind of IP threats have already occurred or could be likely to occur and what type of IP protection means are already in place.

It can be helpful to systematically track the kind of infringements that are occurring and to document them on a continuous basis. At the Technology Center of

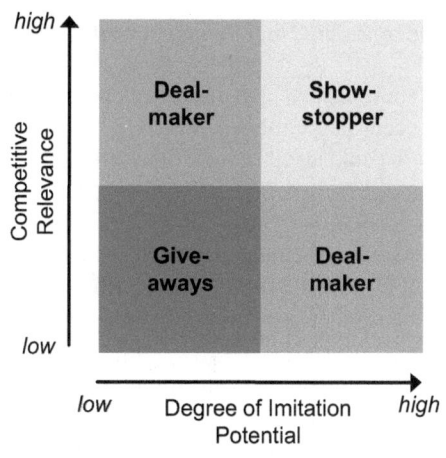

Give-aways
• Available enabling technology
• No differentiation towards competitors

Deal-makers
• Know-how to be specifically evaluated
• Crucial question: How much
  competitive advantage would be lost?

Show-stopper
• Long-term competitive advantage
• Loss contains risk of building up
  competitors quickly

Fig. 5.2 Siemens China know-how evaluation matrix

Siemens China the technological know-how is evaluated according to two criteria, that of competitive relevance and the degree of imitation potential (see Fig. 5.2).

If the competitive relevance is low, the know-how is either state of the art and does not offer differentiation potential or it is difficult to imitate and can become a deal-maker. In the latter case, the know-how needs specific evaluation answering the question of how much of a competitive advantage it can leverage. A similar situation occurs if the degree of imitation potential is low and competitive relevance is high. In the case that the know-how offers a high competitive advantage and the degree of imitation is high, the loss or misuse of know-how bears the risk of strengthening competition. That kind of know-how is a show-stopper. It is of proprietary nature and secures long-term competitive advantages. It should not be shared with third parties. Before entering cooperation, the know-how evaluation helps the R&D managers to determine how to share the different types of information. Another European firm identifies and evaluates their key technologies and know-how on an annual basis with a clearly delineated definition of the competitive relevance. The strongest dimension determines if the technology is or can be patent protected within the next 3 years. The next dimension considers a technological advantage, which allows a lead-time of at least 1 year to the strongest competitor. An equal positioning exists if the strongest competitor has the technological potential to be 1 year ahead in time or may fall back 1 year within the next 3 years. The development managers review the evaluation and discuss the underlying projects once a year. Furthermore, the technological differentiation in addition to cost and quality differentiation potentials is evaluated for key technologies and relevant know-how. The evaluation outlines the legal protection by patents and the factual protection by technological differentiation of the firm's key technologies. The evaluation of know-how and its protection requires two fundamental distinctions of knowledge: its tacitness and its codification. Since know-how drainage is embedded in the individual action and illicit use of know-how, the sources of knowledge and the knowledge characteristics are essential for know-how protection.

> **The following question checklist should be considered in the evaluation of technologies and know-how:**
> - What kind of IP threats (legal-illicit, -licit, factual) have already occurred against the firm or could be likely to occur in the future?
> - Which key technologies and know-how could attract competitors or copyists? Describe competitive relevance.
> - Which key technologies and know-how are easy to copy? Why?
> - What kind of differentiation to competitors can be realized by the key technologies and know-how? What kind of customer benefit can be realized?
> - Can we protect the differentiation and customer benefit by patents?
> - What kinds of patents exist already?
> - Which other legal-, human-, technology-, business-, and market-driven protection means protect the differentiation and customer benefit?
> - What would be the damage if the key technology and know-how would be copied?

5.2 Composition of Preventive Protection

The application of IPRs typically protects a certain technology, technical feature or design. The fact that a product innovation often consists of several technical inventions leads to the assumption that a product is most likely protected by multiple IPRs. The patent protection of single inventions only safeguards the technical invention as such. This results in the protection of mutually exclusive IPRs for the product innovation. The actual customer benefit and competitive advantage are not necessarily protected. The integrated protection by patents, trademarks, utility models and factual protection means offer a greater potential for protection. The protection of the competitive advantage by IPRs and their specific and synergistic combination supports the protection of the perceived customer value and gives rise to a quasi-monopolistic advantage over the competition.

Google's business model is based on the provision of advertising space with a customer specific allocation and display of adverts. Although the search engine is the actual Google product, the business model is based on the customized display of adverts. Google protects the underlying algorithm.

> **The preventive protection wall shall consider the defense minimum level. Following questions shall identify the preventive protection wall:**
> - What is the value proposition to the customer?
> - What are the main differentiation elements?
> - Can we protect differentiation and customer benefit by patents?
> - What kinds of patents exist already?
> - Which other legal, human, technology, business, and market-driven means protect differentiation and customer benefit?
> - What would be the damage if key technology and know-how will be copied?

Case Example: Fischer Wall Plugs

Fischer wall plugs are internationally known to be high quality products. The company is constantly innovating to make their products more durable and more reliable. Given the fact that many of their wall plugs are made out of plastic they are often targeted by Chinese imitators.

In 2010 for instance, a German dollar store sold wall mounting sets consisting of screws and wall plugs. The wall plugs were an exact copy of the patent protected fischer UX wall plugs. The fischer wall plug is protected by two patents, both were infringed. Lab tests showed that the plastic used for the counterfeited product was of a far lesser quality.

Fischer came to an agreement with the German dollar store.The store handed over all wall mounting sets as well as the name and the contact information of the Chinese producer. fischer decided not to go to court in China but to settle the dispute directly with the infringing company. The infringer signed a cease and desist letter and destroyed the molding tools.

Schrauben -und Dübelset groß

1€

Original   Copy

5.3   Piracy Intelligence and IP Monitoring

Piracy intelligence is an activity that enables companies to identify piracy threats and opportunities that effect the future growth and survival of the business. It aims to capture and disseminate required piracy information needed for the evaluation and decision making in individual infringement cases. Due to the local and global distribution of imitations, effective piracy intelligence is becoming increasingly important. Recognizing the importance of this issue, firms implement and develop piracy intelligence processes designed to identify and capture information and evidence about emerging imitations. Visits at trade fairs with particular emphasis on searching for infringements, internet searches and regular scanning of industry specific magazines, competitor's product information are only one instrument to identify IP threats in or from China.

A common method to discover patent infringements is the re-engineering of products. It renders insights into competitive technologies as well as their application. Since R&D personnel usually fulfill re-engineering, their knowledge about the IP portfolio is the basis to identify infringements through re-engineering. Furthermore, the sales personnel that are in contact with customers need this knowledge to be capable to identify know-how theft and IP infringements. Fundamentals of IP knowledge are necessary in order to educate the customer about possible infringements. For the identification of know-how theft and the collection of proof of evidence, professional investigators are recommended. However, the right choice of a professional agency, which is both, loyal and professional, can be challenging. The recommendation of international firms is helpful since the selection of an investigator without any references bears the risk of disloyal investigators.

European firms have already suffered from devised and fake proof of evidence of such investigators. The professionalism, loyalty and reliability of investigators as well as their local networks are important since they need to get access to the infringer's information base directly or via accomplices. Stolen codified knowledge on data storage devices such as DVDs, CDs, USB storage devices, or paper might be found at the facilities. Another less risky way is to request offers with technical details, which may render indications for a theft of know-how. By requesting specifications, which can only be provided by the proprietary technology and know-how, the theft of proprietary know-how can be identified by demanding the relevant documentation as proof of competence. Products or spare parts can be acquired by means of test purchases and then tested and evaluated by an authorized examiner. It is important to assign an authorized examiner who can write a notarized report about the fake parts as evidence. Due to the obligatory notarization of the proof of evidence, the authorized examiner would have to act as a purchaser under a fake identity requesting the product, technology or know-how. The technical discussions and the presentation of proprietary, copied documents or parts should be video or audio taped to use it as proof if necessary.

The cooperation with an agency and an authorized examiner allows a more efficient collection of evidence. The notarization can be realized and allows bringing forward all notarized evidence, e.g., pictures, video or audio recording, witness statements as well as reports and product parts. In the case of a successful proceeding, the infringer will have to stop his infringement and might have to pay compensation fees. Since the enforcement of such fees is critical due to changing company names and addresses, the forbearance, stopping and deterrence of further infringements are often the main results of such activities. An active communication with customers about the involvement of the infringing company in a lawsuit in China may yield to less sales of the infringer and revitalize own sales.

Another instrument to keep track of IP threats in China is the monitoring of Chinese IPRs. Due to language differences the monitoring of IPRs requires a systematic approach to reduce the efforts for an IP monitoring. The selection of key words as well as the main competitors can be used as a first filter to identify relevant IPRs. The scanning of the English abstracts renders insights if a translation is considered for further information. Due to the former absolute novelty definition, Chinese utility models might be based on state-of-the-art outside of China. With the new amendments of the patent law, which came into force in October 2009, this legal gap has been closed. Thus, the risk of infringing third parties' utility models due to the use of state-of-the-art technologies is diminished. Also, the risk that patented technologies (outside of China) are applied as utility models should not occur any longer. The monitoring and scanning of IP also allows identifying new entrants in the market. Particularly in China, the vast geography may not allow an identification of a competitor before the publication of an IPR. The monitoring of IPRs can facilitate its identification.

Main questions to consider for Piracy Intelligence and IP Monitoring:
- What kind of infringements and IP threats are endangering the own business in or outside of China?
- Are external investigation agencies needed for supporting the identification of infringements?
- Which key technologies and know-how need specific monitoring?
- Do own dealers, sales personnel, employees, customers, and their employees behave loyally?
- Which regions need special consideration?

5.4 Evaluation of Infringements and IP Threats

Identified results of IP threats from the piracy intelligence and IP monitoring have to be evaluated to decide on relevant actions. The IP threat typologies help to cluster the threats into legal-illicit, legal-licit and factual threats. Each case has to be considered individually to determine appropriate actions. For the evaluation, a network diagram or a decision tree is recommended (Fuchs 2006; Chen 2010).

The limitation of a decision tree is the requirement of knowing the IP portfolio. As explained, different employees or customers may discover IP threats and need the awareness and knowledge about the companies' proprietary rights. Or, if unsure about an infringement, they need at least the knowledge about communication channels. They should be able to report directly and ad hoc to the responsible persons for IP protection and enforcement. Once the information about an IP threat is received, the decision tree starts with the evaluation of the infringement. An example of such a decision tree can be found in the appendix (Decision Tree—Ciba).

At Ciba, once this information is gathered, a mutual decision about the reactive measures is discussed with the product management and the IP department. The main question in this discussion is: How much impact has the IP threat on the business? The implications to own products and technologies are evaluated and the loss of customers or a reduced differentiation has to be determined. Also, the existing proof of evidence needs to be considered in the decision-making. Usually, infringed parties should try to resolve IP threats outside of court and might start with a cease-and-desist letter. In the case of a positive reaction of the infringer, the halt of violation shall be monitored. In case no action takes place, a second warning letter shall demand the stopping of fraudulent actions. Also a direct confrontation may help to negotiate about a halt of violation. If no reaction occurs, a cost-benefit analysis helps to determine whether to enter a juridical track or to tolerate the infringement. However, if the compensation damages are high and can be determined, a broad and notarized proof of evidence exists, the infringer is not a state-owned company and the venue of jurisdiction and the authorities are neutral to the infringer, litigation might be entered instantly. The own external guanxi can help to support the evaluation and reaction process.

A difficulty in practice is often that the compensation damages are hard to evaluate or do not render the desired amount. According to the Chinese patent law, there are three different methods on how to determine the compensation income for damages caused by an infringement: (1) On the basis of the losses suffered by the patentee or (2) the profits which the infringer has earned through the infringement or (3) if it is difficult to determine the losses which the patentee has suffered from or the profits which the infringer has earned, the amount may be assessed by reference to the appropriate multiple of the amount of the exploitation fee of that patent under a contractual license. But, if there are no illegal earnings, a statutory fine can be imposed which is limited to the amount of RMB 1,000,000 Yuan.

In practice these options are often found to be insufficient, however, due to the limited options the licensing analogy is common practice. Since most cases are solved out of court, the alternative protection means should complement the legal protection. The identification of the source of infringement should be the objective to sustainably protect and deter further threats. Depending on the case and the perceived imitation, the customer may need to be educated about possible imitations and the differentiation of the own technologies and products. The application of further protection means will be of reactive as well as preventive character.

> **Main questions to consider for the evaluation of infringements and IP threats:**
> - What kind of IP threat exists?
> - What kind of business impact does the infringement have?
> - What/who is the source of infringement?
> - What are the governmental relations of the infringer?
> - Does own proprietary IP exist and is the relevant country covered?
> - What kind of evidence exists or would be necessary?
> - Can a cease-and-desist letter deter that infringer?

5.5   Application of Factual and Legal Protection Means

Many companies start their IP protection once an IP infringement has already oc-curred. In case the infringement would not allow high compensation damages, liti-gation is usually not initiated. Also, if the probability of success in litigation is low (e.g., due to limited proof of evidence or expected local protectionism) litigation is not entered. However, the administrative proceeding allows the forbearance and stop of infringement. Thus, the most common way to counteract IP infringements is the administrative proceeding to forbid the fraudulent actions. Traditionally, con-flicts with low expectation of litigation success are solved without a conflicting legal proceeding. This also applies, if it can be anticipated that an infringer will stop the fraudulent action once educated about the infringement and deterred by the menace of a legal proceeding.

From a legal-contractual point of view, it is common practice to include an ar-bitration clause in contracts. Particularly in China, the tradition of not loosing face and to respect the counterpart is an important cultural aspect that needs consider-ation. Negotiations can be useful if the infringer is a partner, supplier or customer with whom further relations are favored. Particularly if the infringer might be a potential trading partner, negotiations are important. Some Chinese producers have a limited knowledge about intellectual property protection and legal consequences. They might test the reaction of a genuine producer and will stop their activities consequently after a cease-and-desist-letter.

In case the infringement of an IPR may render high compensation, it might be worthwhile to consider litigation. The expected damages for IPR infringement are one of the key factors to start infringement litigation. So is the question of cover-ing the costs of litigation. As those two issues, damages and cost coverage in in-fringement litigation, reflect the degree of protection through the legal system, these aspects of IPR litigation play a significant role for foreign firms when deciding whether or not to litigate (Hammel 2006).

China is following the 'equity doctrine'. That means it deems at bringing the IPR holder back to the state before its rights were infringed.[1] In case it is not possible or

---

[1] This evaluation approach is reflected in Art. 56 PRC Trademark Law, Art. 20 PRC Unfair Com-petition Law and Art. 48 PRC Copyright Law.

difficult to determine the losses of the IPR holder, the profit of the infringer or the royalty fees for a patent, the amount of compensation damages is provided by statutory damages. The Chinese courts are entitled to calculate these damages depending on the circumstances of the infringement but only up to an amount of RMB 1 million Yuan.[2] This amount does not limit the amount of an exact evaluation and calculation of damages; these damages may exceed the statutory damages amount by far.

For IP threats without a legal infringement, preventive measures should be considered. Also an intensified IP piracy and IP monitoring might be helpful to detect a threat at an earlier stage (for both see Chap. 4). The first reactions to an IP threat should be preventive countermeasures consisting of legal and factual protection to reduce the risk of further or additional drawbacks.

For the evaluation of an infringement and its opportunities of success in litigation a professional local legal advisory should be considered. The decision on how to counteract a threat should be realized on a case-by-case basis. A common rule on how to legally and factually counteract infringements is not purposeful.

Main factors that determine the application of legal and factual protection means are the probability of success, the compensations versus costs and the short- as well as long-term impact of protection.

> **Main questions to consider for the application of factual and legal protection means:**
> - What kind of infringements and IP threats are endangering the own business in or outside of China?
> - Is protected IP concerned? What would be the probability of success?
> - Which dimension offer opportunities to prevent or counteract such infringements?
> - Which factual legal protection means exist to prevent and counter such infringements?
> - Which protection dimensions are most relevant?

5.6 Institutionalizing IP Protection

Managing the IP protection in China effectively requires a continuous control and improvement process. Direct and indirect protection means should be used for a systematic IP protection process. In practice, the IP protection approaches have been organized as a workflow organization (Fuchs 2006). Vital to the successful protection is the application of different IP protection means for prevention, identification and enforcement. The ownership of different IP-relevant tasks is essential to create change. Different roles include mainly the IP and legal department but also

---

[2] This limit has been increased in the revised PRC Patent Law coming into effect in October, 2009 (Art. 65, Paragraph 2). The former amount was limited to 500,000 Yuan RMB.

the top management, product management, product development, human resources, marketing, corporate communication, sales and production.

For safeguarding innovations from IP threats a clear protection strategy and process as described in the previous chapters is fundamental. The responsibilities start with the establishment of an IP strategy, which creates value complementary to protection and puts forth the competitive advantages by means of proprietary IP. The top management has to ensure that throughout the value chain the responsibilities are set. In case of a serious IP infringement the top management has to get involved in order to decide upon litigation. The product development has the responsibility of developing differentiating features and to protect that differentiation by means of intellectual property rights. The elaboration of critical know-how and what to share in collaborations with partners is essential. The application of IP may support such collaborations due to the proprietary nature of IP, which may allow a temporary monopoly. The application of IPRs is a preventive protection that allows the protection of differentiation. Furthermore, preventive measures such as conceptual or technology protection means increase the barrier for imitation. The application of both, the black-box principle and the need-to-know principal are means, which reduce the risk of know-how drain. The reengineering of third parties' products allows the identification of IP infringements provided that the researchers and developers know the company's IP portfolio.

A distributed production reduces the risk of a single person knowing the entire value creation and process know-how of the production. Restricted physical access control to the production sites is only one of the instruments to increase the drainage of know-how. The product management has the responsibility to evaluate the imitation attractiveness, which helps to decide on application of preventive protection means. In cooperation with the marketing, they decide on brand and service differentiation. In case of an IP infringement, the product management has to determine the business impact as an input for the IP department. In serious cases they work hand in hand to prepare a decision request for the top management. The human resources need to ensure that critical know-how profiles are set up. The systematic recruiting and staffing of loyal employees for critical profiles is vital to prevent know-how theft. The critical profiles shall also include the rules for physical and logical access rights. Also the motivation and a transparent incentive system (e.g., included in a complementary IP policy to the employment contract) help to sensitize the individual employee. The creation of an internal guanxi is crucial to retain employees and to exploit their external guanxi (Figs. 5.3 and 5.4).

Particularly the sales personnel with a broad external guanxi towards customers can be beneficial for returning profits from innovation. From an IP protection perspective their guanxi may provide insights into competitive products or offers from competitors. Particularly in the industrial goods industry the identification of IP infringements is not observable at first sight. Thus, the guanxi may support piracy intelligence for an early warning of potential infringers. The corporate communication supports by communicating an active piracy intelligence and IP enforcement.

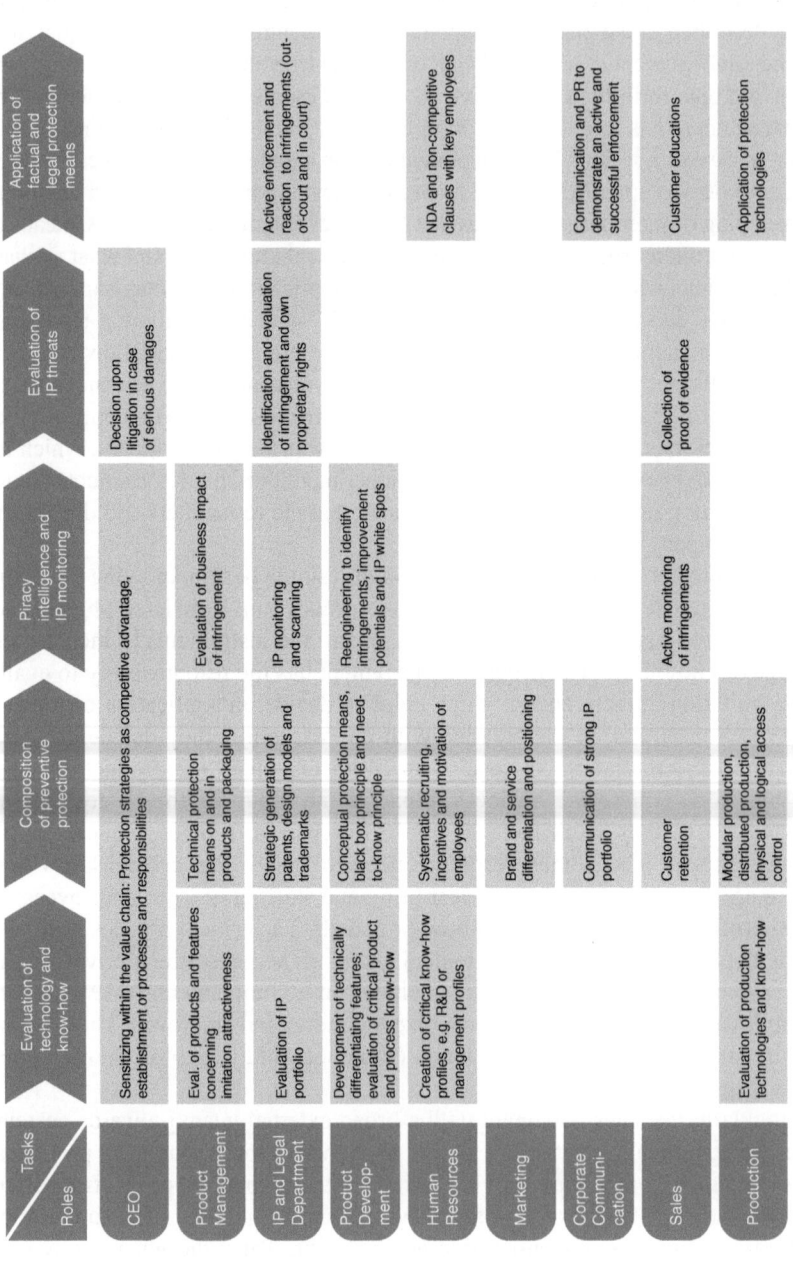

Fig. 5.3 Beyond the IP department: Roles for IP protection

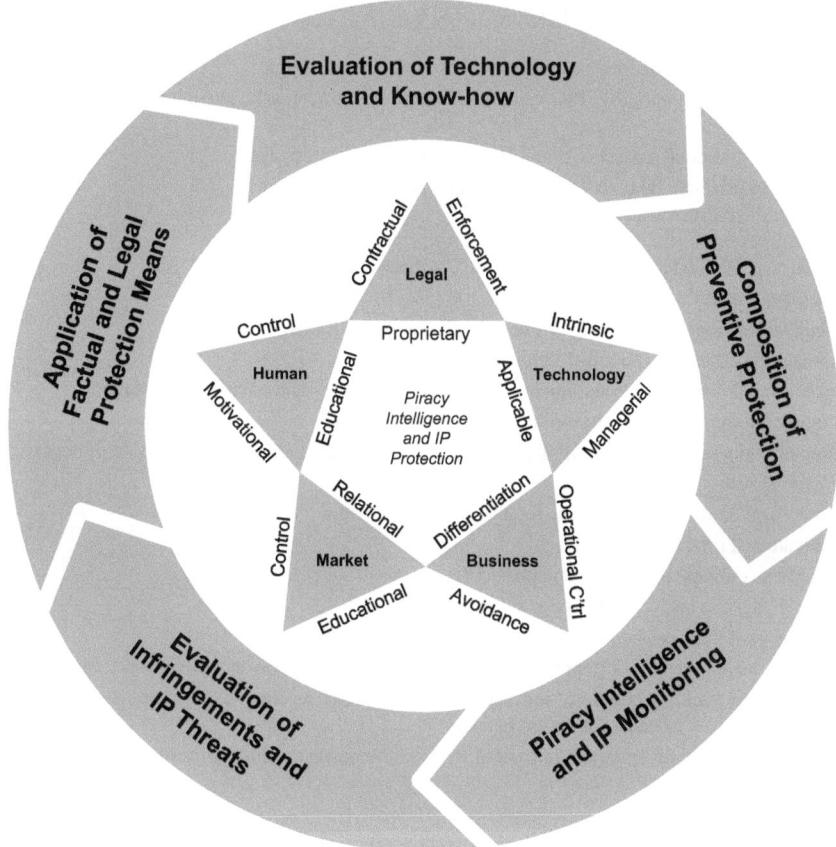

Fig. 5.4

5.7   The Integrated Management Model for Piracy Intelligence

The continual improvement of piracy intelligence and IP protection needs a clear strategic direction and task ownership. Since IPR infringements are only part of the IP threats in China, a typology of IP threats is presented. In addition to legal-illicit threats, legal-licit threats such as uncertainties within the boundaries of the law and IP-related factual challenges are also predominant in China. Different dimensions of protection means incorporate technology-driven, human- and business-related as well as market-driven protection. The following table summarizes recommendations for their application.

In order to pursue a continuous IP protection policy through factual and legal protection, the responsibilities for tasks are distributed within the organization. The ownership of piracy intelligence and IP protection tasks is vital to effectively apply

Table 5.1 Recommendations for application of IP protection means

Legal-driven protection
Apply IPRs as a baseline for protection
Protect value proposition by IPRs using combinations of trademark, patent and design model protection
Agree on contractual agreements and IP policy with key employees
Enforce proprietary rights to deter existing and potential infringers • Select the correct venue • Assign external support for the collection of evidence • Use a specialized external IP agency for IP enforcement
Market-driven protection
Establish relationships with governmental officials and customs
Educate the customer about differentiation and IP
Exploit relationships with customers and partners to identify and counteract IP threats
Monitor the market to identify imitation and other IP threats
Monitor Chinese IPRs to identify legal threats, new entrants and to ensure freedom of action
Human-driven protection
Select loyal employees and create loyalty and IP awareness
Motivate and retain key employees
Provide incentives to generate and protect IP
Control physical and logical access to key information
Technology-driven protection
Evaluate the application of visible, invisible and/or machine-supported protection technologies
Consider and exploit technology-intrinsic mechanisms such as complexity, process intensity and specialization to create imitation barriers
Increase lead-time advantage by means of R&D in the newest technologies or its transfer to China
Business-driven protection
Increase imitation barriers by means of differentiation
Identify key information and know-how to decide on avoidance tactics: keep key information secret, keep R&D in home country or limit for know-how distribution
Outperform potential imitators by process excellence and cost leadership
Deter and counteract IP threats by operational control

protection means. Deliberate communication and contact persons allow an ad hoc reaction to threats.

In order to ensure an effective learning process and enhancement of piracy intelligence and IP protection a continuous improvement process must be established to prevent and counteract IP threats in the dynamic Chinese environment (Table. 5.1).

Using the IP Protection Star Effectively

6

The perspective of IP protection on appropriability mechanisms enlarges the concept of appropriating returns on innovation by creating and safeguarding value from IP protection through different means. Furthermore, the introduced IP threats, namely legal-illicit, legal-licit and other factual threats, offers evidence that the appropriability regime in China is not only afflicted with a comparatively weak enforcement of intellectual property rights but also with further legal-licit threats that comprise uncertainties and challenges within the boundaries of the law. Other factual threats, which are predominant in China support the weak appropriation of returns from innovations. The identification of different threats beyond the legal dimension supports the demand for factual protection.

The protection means, technology, human, market and business are vital to embrace the dynamics of IP threats in China. The piracy intelligence and identification of IP infringements is challenging. Due to the dominant view that IP professionals autonomously manage IP activities, the traditional IP protection is mainly characterized by the application and management of proprietary rights. However, one has to acknowledge the emerging principle that IP protection creates value, which, in itself, creates protection. The protection of the value proposition creates differentiation to competitors, which in turn creates the appropriation of returns from innovations. Institutionalizing IP protection creates change and comprises task ownerships with clearly defined responsibilities (Table. 6.1).

In order to counteract threats, an IP protection demands specific protection means selected on a case-by-case basis. The pervasiveness of the view that IP protection mostly consists of legal protection is incomplete. The refinement of legal and factual protection means into a human-, market- and business- as well as legally- and technology-driven protection leads to a comprehensive view on IP protection options. The integrated management model for piracy intelligence and IP protection follows a five-step approach on how to identify IP threats at an early stage and how to apply different protection means. IP tasks for different roles should be created by ownership to obtain change and improvement in IP protection (Fig. 6.1).

O. Gassmann et al., *Profiting from Innovation in China*,
DOI 10.1007/978-3-642-30592-4_6, © Springer-Verlag Berlin Heidelberg 2012

Table 6.1 IP threat typology

Legal-illicit threats	Legal-licit threats	Factual IP-related challenges
Invention patent infringement	Circumvention of protected technologies and products	Drainage of know-how and unprotected IP
Trademark infringements	Vast amount of invention and utility model patents in Chinese language only	Cultural differences and lack of IP awareness
Design patent infringements	Former definition of novelty in China	Local protectionism and privileges
Copyright infringements	Obscurity of interpretations of legal changes	Demand of disclosure of know-how for potential project acquisition
Trade secret infringements	Uncertainty of success in litigation	Copy of unprotected technologies or process know-how
Own risk of unknowingly infringing third parties' IPRs	Costs vs. restricted compensations for litigation success in China	Usage of the positive connotation without IPR infringement
Contractual breaches by suppliers and partners	Proof of evidence for infringements	Competitive pricing and second mover advantages
Obscurity of criminal networks and logistic chains		
Parallel trade		

Based on our experience we provide the following recommendations. This summary covers the implications for management practice to improve the IP protection in China by legal and factual means that help to identify, prevent and encounter IP threats.

6.1   Determine What Needs to be Protected

The evaluation of technology and know-how provides the basis to determine what needs to be protected. Additionally, a risk assessment should be conducted to evaluate potential hazards and their impact on business so as to best decide on how to protect intellectual property. The evaluation of technologies and know-how provides initial guidance on which protection means are appropriate. In the case where risks outweigh the business opportunities, an avoidance tactic is recommended. In this instance, key corporate know-how and technologies should remain in the home country or within strong appropriability regimes. '

6.2   Be Creative in IPR Applications

In order to establish a thorough defense position in China, a broad and diversified IPR portfolio is essential. Trademark, patent and design model protection are grounds for legal actions. Despite the weak enforcement regime, IPRs are neces-

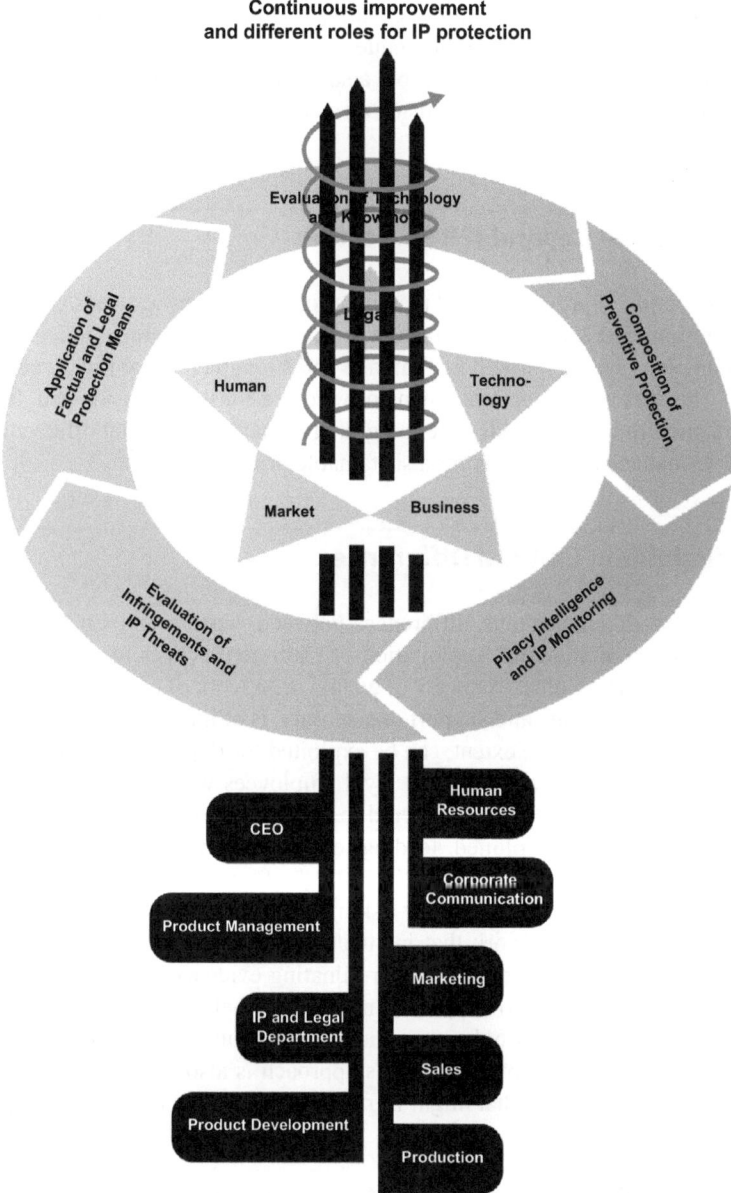

Fig. 6.1 Integrated management model and roles for continuous improvement of piracy intelligence and IP protection

sary to protect the value proposition. Creative combinations of IPR and trademark protection should consider the entire value proposition for the customer and should include not only the key technologies but also customer benefits such as ergonomics or services. Strong defense mechanisms also strengthen the bargaining position in the case of licensing agreements.

6.3   Protect IP Beyond IPR Protection

The introduction of preventive and reactive protection means exceeds by far the legal dimension. The application of factual protection in complement to legal protection means can prevent know-how drainage and strengthen the ability to counter IP threats. The different protection dimensions require constant revision and, if necessary, adaption to current needs. The integrated management model offers guidance on how to manage these means for a sustainable protection.

6.4   Assimilate Cultural Differences

The existence of fundamental differences between western and Chinese cultures demands an assimilation and exploitation of Chinese practices in order to achieve positive results. The Chinese concept of guanxi, a network of relationships whereby various parties cooperate and support one another, is of high relevance for business opportunities and, to some extent, can be exploited for IP protection. The establishment of internal guanxi gives rise to loyal employees with a strong commitment to the firm. If an employee is discovered to be involved in the theft of know-how, internal guanxi can be exploited, leading to the guilty party's loss of face within the network. Although ethically questionable, the fear of losing face and an exclusion from the network can reduce the risk of know-how drainage. External guanxi describes the external network that includes governmental officials and customs, which is helpful for the selection of incriminating evidence or the enforcement of IPRs. Close relationships to suppliers and partners help to increase their loyalty. Guanxi with customers not only strengthens their loyalty but also helps to identify possible competitors and infringers. This approach is also helpful to get access to products (e.g., machinery) that might infringe one's proprietary rights.

6.5   Create Awareness, Retain Key Employees and Exploit Advantages as a Foreign Firm

The retention of key employees is necessary to reduce the risk of know-how drainage. The creation of incentive systems to generate IP and to foster IP awareness should be implemented unconditionally. An IP policy similar to the example presented (in Chap. 4) can be implemented for all R&D employees as a complement to the employment contract. The motivation and incentives to stay within a firm need assimilation into Chinese practice. So-called welfare packages are common practice with state-owned firms in China. To win and retain the best talent, such welfare packages (for example rent allowances, provision of housing or vocational training of the children of key employees) are vital for a firm to become an attractive and competitive employer in China. As an international company, prospects for international promotion should be exploited to win and retain key employees.

6.6   Establish Early Warning Signals by Means of Piracy Intelligence and IP Monitoring

The early identification of possible IP threats is necessary to impede imitations or other IP violations. The establishment of clear communication channels with the IP department in case of identified threats allows an ad hoc reaction. Early warning signals such as information received from customers or suppliers about copies or know-how drain, imitations at trade shows, publications of trademarks or proprietary know-how in or outside of China need to be identified. A continuous monitoring of competitors, publications, websites and IP violations at trade fairs are just some of the means to identify infringements. Moreover, building close relationships with customers and customs (external guanxi) in order to obtain their insights on developments of competitive products and services can provide useful information. Furthermore, the monitoring of Chinese IPRs is necessary to remain free of third parties' rights. It can also help to understand the bargaining position of a competitor or infringer and thus, how to approach negotiations when entering licensing agreements.

6.7   Generate IP for Differentiation and Exploit Differentiation for Protection

A multifaceted IPR protection policy should protect the value proposition and thus the differentiation of one's products and services. Concurrently, the development of differentiation often creates advantages that inherently protect a firm's product offering. We suggest that brand differentiation, service differentiation, price and reliability differentiation support the appropriation of returns and indirectly protect innovation. Innovation speed, strategic release planning and competitive product lifecycle management strengthen the competitive advantage of a firm in China. Business model differentiation, for example a direct sales force in combination with service differentiation, raises barriers to imitation. Due to the multifaceted possibilities to create differentiation, the mentioned list of protection by differentiation remains incomplete but puts forth examples exploited in practice.

6.8   Outperform Imitation by Process Excellence and Operational Control

Competitive pricing is crucial in the Chinese marketplace. A competitive pricing policy is necessary, since the price is often the main reason behind the popularity of copied goods for customers. Cost-effective production processes are required to achieve that objective. However, the transfer of production know-how needs careful consideration and, if the risk of losing core competencies outweighs the business benefits, the production technology should remain in the home country. In addition to a competitive production process, the operational control of the flow of components and products has to ensure a secure supply chain.

6.9   Exploit IP Threats and Learn from Imitation

Considering an imitator as a second-mover competitor, imitations can be considered as learning opportunities for the original product development. Reengineering of imitations makes the differences in product and production technologies accessible. Complementary to the technical learning from imitations, such competition may also give deeper insights into customer needs. In particular, the reasoning behind customers' decisions to deliberately buy or refuse to buy an imitation can be crucial for customer-driven innovation. Learning from imitations and the customer's rationale for buying imitations needs to be an integral part of a continuous IP protection policy.

6.10   Define Protection Tasks and Roles

IP protection means are manifold and their application has to be assured by a number of persons. The IP and legal departments are responsible for the legal protection of technologies and services and have the role of a custodian for IP protection management. With top management support, the different tasks and roles within the organization have to be defined and implemented. It is crucial to sensitize middle management to the importance of IP protection since IP threats are often identified at/or reported to that level. The implementation of factual protection means has to be integral to all mentioned departments.

6.11   Establish Task Ownerships to Create Change

The ownership of different tasks and the commitment to the underlying role creates change. For a continuous improvement of IP protection over time, it is vital to obtain regular reports or statements on the topic of IP threats and IP protection. Without the demand for such information from the top management, the responsibilities often remain solely at the IP department level.

6.12   Reduce Risk and Safeguard Business Opportunities by Avoidance or Tolerance

In case protection cannot be guaranteed and the negative business impact of infringement or IP threats outweigh the opportunities, the avoidance or toleration of IP threats is a final option. Overall, the situation in China remains challenging for IP protection and enforcement. If the IP threats create serious damage to the firm, a final protection strategy involves withdrawing or staying out of China altogether. In cases where only certain product lines are affected, the market entry in weak appropriability regimes might follow a 'waterfall strategy' or strategic release planning: Once the product has been introduced in the home country, it will be subsequently introduced to China at a later point in time.

5.12   Reduce Risk and Safeguard Business Opportunities by Avoidance or Tolerance

Outlook

Henkel's CTO once ironically told us: "Don't bother about patents in China. It only restricts our strategy for the next 20 years." China is making a concerted effort to build a more innovative economy by heavily investing in R&D and education. Concurrently, the country has made tremendous progress in developing an IPR system since the mid 1980s. However, the protection of proprietary IP in China remains a challenging task. It will require several more years and a revolutionary rethink before China develops into a strong appropriability regime. The outline of a national IP strategy was introduced in 2008 to improve the ability to administer, protect and exploit IP. With the revision of the patent law, which has been effective since October 2009, some legal gaps have been closed. Under the former patent law, prior use of an invention outside of China (except for public disclosures) did not defeat the novelty of the invention. Thus, there was a constant risk that any Chinese patent and utility applications would be understood as absolute novelty. Under the revised patent law the territorial limit has been removed and lifts the standard to an absolute novelty definition comprising state of the art known to the public within and outside of China before the publication date. With this and other changes harmonization with worldwide standards has been further strengthened.

The increasing number of patent applications in China by domestic firms underlines the importance of IP generation by local firms to protect their innovation portfolios. The Chinese government continually invests in innovation and has strategically placed it at the center of its 5-year development plan. As Chinese firms mature, innovate, and develop their own IP they will naturally demand a better IP system. With high export rates, Chinese firms are already confronted with IP protection outside of China. The development of indigenous innovations will strengthen the IP protection system. Furthermore, growing awareness of the existence of IP among the general population, made possible through the rising standards of education will provide for a better overall understanding of IP issues. With ever-increasing numbers of innovations and IP breakthroughs being made by local firms, Chinese managers will naturally seek to better secure them, looking to their governmental

O. Gassmann et al., *Profiting from Innovation in China*,
DOI 10.1007/978-3-642-30592-4_7, © Springer-Verlag Berlin Heidelberg 2012

structures and legal systems for support. However, as long as the Chinese IPR regulations and enforcement remain weak, IP protection will remain one of the highest priorities for industrial firms in China. If China continues its innovation efforts, whilst also reinforcing its intellectual property system, it will become an innovation powerhouse within the next two decades.

Appendix

Requirements for Non-PCT Chinese National Phase Application

The requirements for the non-PCT application are similar to the PCT Chinese national phase application, however, they differ in terms of the timeline, priority claims and amendments.

1. One copy of the specifications of the application including abstract, description, claims, and drawings.
2. Information of priority if priority is claimed. If priority is claimed, the priority document has to be filed within 3 months from the filing date.
3. Indication if the request for substantive examination should be filed simultaneously with the application.
4. Amendments to be made at the time of filing.
5. The name of the applicant in Chinese characters. Also, the applicant's name and address in English.
6. The name of the inventor(s) in English and if the person is Chinese also in Chinese characters.
7. Indication if the request for substantive examination should be filed simultaneously with the application.
8. Indication if the re-registration in Hong Kong should be filed.
9. Power of attorney. Form can be handed in within 3 months of the filing date.
10. In case the applicant in the priority document is different from the applicant in the Chinese application, an assignment document is required. The document can be in English as a certified copy from the patent office or as a notarized copy of a notary or the original copy with original signature.

O. Gassmann, *Profiting from Innovation in China,*
DOI 10.1007/978-3-642-30592-4, © Springer-Verlag Berlin Heidelberg 2012

Requirements for the Application of Design Patents

For the application of design patents the application should comprise the following documents:

- Letter of request. The letter of request for a design patent specifies the product incorporating the design and the class to which the product belongs to.
- Brief description of the design. The description explains important details of the design.
- Drawings or photographs of the design. The pictures illustrate the protectable design. The size should be no smaller than 3×8 cm and no larger than 15×22 cm.
- Prototype or model of product. Where necessary, the product that incorporates the design should be submitted.

Changes of the Revised Chinese Patent Law in Comparison to Former Law

Extracted from Wong, K. and Low, E. (2010): China: Intellectual Property Law in the People's Republic of China: An Update. JSM, Mayer Brown International LLP.

On October 1st 2009, the revised patent law of China came into effect. This law has resulted in significant changes to China's patent system, from the preliminary stage of patent application to enforcement of patents in the courts. Some key changes in the new patent law are:

Higher Standard of Novelty

In an application for an invention patent or utility model under the old patent law, prior use of the invention outside China (except for disclosure in a publication) did not defeat the novelty of the invention. China's revised patent law removes this territorial limit and lifts the standard of novelty by stipulating that the applied-for patent must not belong to the state of the art known to the public inside or outside China before the application date.

Foreign Patent Applications

Under prior law, foreign companies wishing to apply for and maintain patents in China were required to engage patent representatives from among a limited class of agents designated by the State Council. The revised patent law allows all patent agents in China to handle foreign-related patent applications. Consequently, foreign applicants can now choose from a significantly larger pool of patent agents to handle their patent matters in China.

First-filing Requirement

Under the old law, Chinese entities (including individuals) wishing to apply for foreign patents for their inventions developed in China were required to make their first patent application filing in China. The revised patent law removes this first-filing-in-China requirement. At the same time, however, the revised law imposes a new measure requiring that all local and foreign entities wishing to apply for foreign patents for inventions or utility models completed in China must first apply to the State Patent Office for a confidentiality scrutiny. If this process is not followed, the subject invention or utility model will not be granted a Chinese patent.

Statutory Damages

The new law codifies and increases the range of statutory damages applicable in certain infringement situations. For example, where the rightful owner's loss or the infringer's illicit profit is difficult to quantify, the court may take into account factors, such as the type of patent and the nature of infringement, and it may award damages ranging from RMB 10,000 Yuan to RMB 1,000,000 Yuan. Under previous law, the cap on damages in such situations was RMB 500,000 Yuan.

Trademarks

In April 2009, the Supreme People's Court of China issued an important opinion addressing certain issues related to the adjudication of intellectual property disputes. The opinion, which is applicable at all court levels in China, contains precise judicial guidelines to courts on how they should approach various trademark and unfair competition issues. The Court's guidelines are intended to promote ideals expressed in the National Intellectual Property Strategy announced in 2008. They include:

- Courts need not award damages or account for profits relative to a mark that is registered but not put into actual use (for example, intended for extorting infringement compensation), although injunctions may still be granted against unauthorized use.
- Trade names or their abbreviations that have attained certain market reputation and public recognition through actual trade use are afforded protection against unfair competition.
- An enterprise name that is properly obtained overseas cannot be defended if its use within the People's Republic amounts to trademark infringement and/or unfair competition under Chinese law.

Well-known trademarks represent a category of trademarks that is afforded wider protection under Chinese law. At present, foreign marks or brand names not registered as trademarks in the PRC can receive protection under the nation's trademarks laws only if they qualify as well-known trademarks within China. Given the importance and power of well-known trademarks, this is an area in which the Chinese government strives for consistency in administrative and judicial practice and for a strengthening of protections.

As part of these continuing efforts, the Supreme People's Court of China issued a judicial opinion in May 2009 offering guidance on a number of questions concerning well-known trademarks:

- The circumstances, under which a court may or may not decide if a brand is well-known, are defined. For instance, the court may determine if a mark is well-known in a trademark infringement or unfair competition claim that involves the mark's identity or similarity relative to an enterprise name. On the other hand, if a trademark infringement or an unfair competition claim fails for want of some other statutory criteria, the court may not make a determination.
- Factors to be taken into account when considering whether a trademark is well-known in China include the mark's fame and history of use, the market share of goods bearing the mark, past recognitions, and the extent and geographical scope of associated advertising activities. Supporting evidence in the form of industry rankings, market surveys, valuation reports and the like are recognized as appropriate considerations by the court.
- The Court found that confusion relative to trademark infringement occurs, when the public does not recognize that products bearing a well-known trademark and a mark under complaint come from the same source. Similarly, confusion exists when the public does not understand that a license, association or other like agreement has been arranged between respective traders.

Trade Secrets

Trade secrets are protected in China under the Anti-Unfair Competition Law. To qualify as trade secrets, data must constitute technical or business information that is:

- Not known to the public
- Of practical use and capable of bringing economic benefits to the owner
- Subject to confidentiality measures adopted by the owner

These criteria are defined and clarified in administrative regulations and, increasingly, in judicial interpretations. A significant judicial interpretation was issued by the Supreme People's Court in 2007 to provide important guidance on fundamental concepts such as public domain and confidentiality measures.

For example, the category of data that is not known to the public is interpreted as information not commonly known to, or easily obtained by, relevant members in the trade. The Court's interpretation provides several examples of what constitutes public information, including information that is common knowledge or industrial practice in the relevant sector; information that can be easily obtained without any charge; and information that solely concerns a product's size or structure and that is readily discernible by the relevant public.

Relative to confidentiality measures, courts will consider a number of factors, including the nature of the information, the intention of the information owner, and the form of measure(s) adopted. A number of circumstances, in which confidentiality measures will be found to have been taken by the information owner, are also described. These include the affixing of a confidentiality label on the device that stores the information, the use of a confidential code, and the execution of confidentiality agreements.

The law, however, does not provide blanket protection for trade secrets, and it contains some extremely important limitations. One of such limitation provides that obtaining trade secrets through reverse engineering of publicly available products will not constitute infringement.

Technology Import and Export

In China, the import and export of technology is subject to a control regime that is principally set out in the Regulations of the Administration of Technology Import and Export. It is important to note that these regulations do not simply govern technology businesses. Rather, the term technology import and export is defined very broadly to encompass any kind of patent (or patent application right) assignment or license, confidential know-how, transfer or technology training. Consequently, a large number of transactions in all categories of business have become subject to control.

Under this regime, certain categories of technology are prohibited from import and/or export entirely, while others require prior approval by the Chinese government. From time to time, China publishes a catalogue of prohibited and restricted categories of technology. Violation of these restrictions may attract criminal sanctions.

Technologies that do not fall into prohibited or restricted categories are free for import and export, provided that the relevant contract is properly recorded with the Chinese commerce authorities. In February 2009, the Ministry of Commerce revised its recordal requirements by stipulating a 60-day recordal period for technology import and export contracts after their effective dates. Although the penalty for non-observance is not defined, outsourcing parties must recognize and comply with this requirement, as it represents the basis for handling tax and remittance matters arising out of any import or export activity.

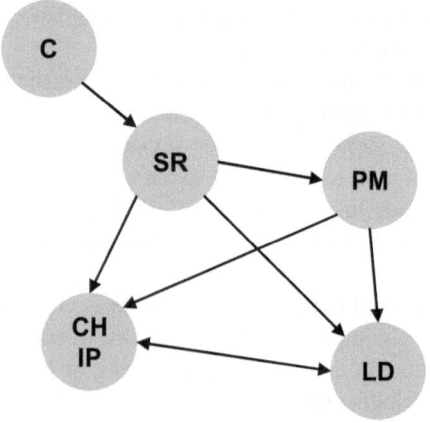

C = Customers that identify IP related   IP = IP manager business unit i
       suspicion   (Switzerland)
SR = Regional sales representative   LD = Law and environment (L&E)
PM = Product manager business unit I   department
                                                  ——▶ = Communication flow

Fig. A.1 Actors of IP threat identification and communication channels at Ciba

Decision Tree—Ciba

> *Even if NDAs[1] are infringed we take actions. No infringement remains without a reaction. We may also sue the person. An attitude like 'I just keep on using the customer list, although I don't work there anymore' is not tolerated. The identification of such cases is easiest via our sales force.*
> *Head of Intellectual Property, Ciba*

The vast geography of China hampers the identification of imitators or IP infringers. So far, Ciba identified imitators or infringements mainly via their sales force and customers. Since the sales employees have access to their customers' sites, they may detect infringements or clarify suspicions, e.g., if a customer suddenly does not purchase at Ciba anymore (Fig. A.1).

The sales force then reports the incident to the product manager or may contact the IP department directly. Generally, sales employees know the corresponding product manager and each product manager knows the IP team in Switzerland. In the case that a customer reports any kind of suspicions, the contacted sales or product manager reports the incident to the IP experts. Short communication flows between product management and IP management allow a fast and direct reporting of threats. Ciba generally does not work with external agents to detect IP threats. However, on a case-by-case basis, the use of agents is helpful. External agents may detect real

---

[1] Non-disclosure agreement.

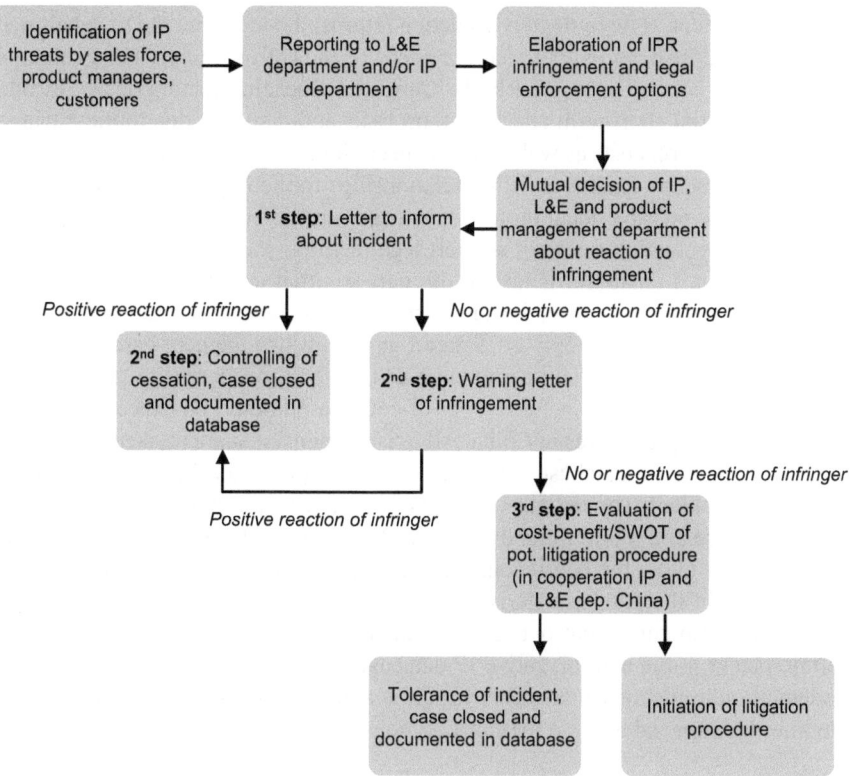

Fig. A.2 Identification and first reactions to IP infringements at Ciba

selling prices and revenues or even purchase an important product. A subsequent analysis of the product will not only offer insights into the competitive product but identify if the product infringes Ciba's IPRs. Trademark infringements are mostly detected via Internet searches. However, IP threat identification is no systematic process; it relies on serendipitous discovery of imitations. Neither a proactive engagement of external agents nor an IP search on trade shows is regularly conducted.

After identification of an IP-related threat, the IP department elaborates what kind of infringement exists and what legal enforcement measures could be undertaken. The legal and the IP managers mutually decide upon the reaction to the incident. The legal department in China usually starts with a letter to inform the potential infringer about the existing problem without any kind of menace. Experiences do exist with such letters to Internet platforms and search engines with a request to abandon the fraudulent use of Ciba's trade name from the platform completely. In most cases the problems can be solved out of court by such correspondence and the infringing brand names can be prohibited and removed from the Internet search platform. If the infringement still remains, the correspondence will be successively tightened. A cease-and-desist letter warns the potential infringer about an

IPR infringement. If no or negative reaction returns, the legal and the IP department evaluate the cost-benefits of an enforcement of the infringement. The evaluation is determined on a case-by-case basis. General influencing factors of that evaluation are potential short-term and long-term risks and damages due to the infringement. Furthermore, costs as well as win expectations due to the burden of proof and contingency factors such as existing relationships to the opponent, the opponent's power, and the possible local protectionism are considered in the evaluation. If the cost-benefit evaluation does not support legal actions, the case will be closed and documented in a database. If no out-of-court solution can be achieved and Ciba resolves on a legal procedure, the administrative way as legal reaction will be initiated. Civil litigations are only considered in patent infringement cases where the damages are high enough and worthwhile a litigation procedure (Fig. A.2).

So far, Ciba has mostly achieved out-of-court resolutions and executed only one or two litigation procedures in China. All infringements that are tracked and reacted to are collected in a database. The database was established in the mid 1980s to capture all incidents concerning IP threats. The legal department acts as custodian of the database and joint efforts of the sales force, product managers as well as the IP department allow the continuous maintenance of the database. The documentation process in the database starts as soon as a letter, a cease-and-desist letter or any other document is sent out. The documentation is not necessarily linked to a certain patent in the administrative IP database. Instead it is part of the patent and trademark administrative database, which is administered separately in a conflict module. This special module for conflicts conserves incidents of all kind.

References

Abernathy WJ, Utterback JM (1978) Patterns of industrial innovation. Technol Rev 80(7):40–47

Arthur WB (1989) Competing technologies, increasing returns, and lock in by historical events. Econ J 116–131

Arundel A (2001) The relative effectiveness of patents and secrecy for appropriation. Res Policy 30(4):611–624

Bader MA (2005) Managing intellectual property in inter-firm R&D collaborations—the case of the service industry sector. Dissertation, Universität St. Gallen, No 3150, St. Gallen

Bader MA (2006a) Intellectual property management in R&D collaborations—the case of the service industry sector. Springer, Heidelberg

Bader MA (2006b) Managing intellectual property in a collaborative environment: learning from IBM. Int J Intellect Prop Manag 1(3):206–225

Behrmann N (1998) Technisches Wissen aus Patenten. Dissertation, Universität St. Gallen, No 2104. St. Gallen

Bjoerkman I, Lu Y (1997) Human resource management practices in foreign invested enterprises in China: What has been learned? In: Stewart S, Carver A (eds) Advances in Chinese industrial studies, vol 5. JAI Press, Greenwich, pp 155–172

Bloch PH, Bush RF, Campbell L (1993) Consumer 'accomplices' in product counterfeiting. J Consum Mark 10(4):27–33

Boisot M, Child J (1988) The iron law of fiefs: bureaucratic failure and the problem of governance in the Chinese economic reforms. Adm Sci Q 33:507–527

Borg EA (2001) Knowledge, information and intellectual property: implications for marketing relationships. Technovation 21(8):515–524

Bosworth D, Yang D (2000) Intellectual property law, technology flow and licensing opportunities in the people's republic of China. Int Bus Rev 9(4):453–477

Boutellier R, Gassmann O, Zedtwitz M von (2008) Managing Global Innovation, Uncovering the Secrets of Future Competitiveness, 3rd revised edition (2nd revised edition in 2000, 1st ed. in 1999), Springer, Berlin, p 629. Chinese Edition: Guangdong Economics: Guangzhou 2002, p 556

Brockhoff K, Ernst H, Hundhausen E (1999) Gains and pains from licensing—patent-portfolios as strategic weapons in the cardiac rhythm management industry. Technovation 19(10):605–614

Browning T (2004) Trade secrets protection in the workplace in China. World Trade Executive, April 2004, pp 16–19

Bush RF, Bloch PH, Dawson S (1989) Remedies for product counterfeiting. Bus Horiz 32(1):59–65

Chen CC (1995) New trends in reward allocation preferences: a Sino-US comparison. Acad Manag J 38:408–428

Chen WC (2010) Protecting intellectual property in developing countries with special focus on China. Dissertation, University of St. Gallen, No 3680, Bamberg

Child J (1994) Management in China during the age of reform. Cambridge University Press, Cambridge

Child J, Möllering G (2003) Contextual confidence and active trust development in the Chinese business environment. Organ Sci 14(1):69–80

Ciba Annual Report (2007) Annual Report. Ciba Inc., Basel. http://www.analist.be/reports/Ciba_2007.pdf

Clark D (2000) IP rights protection will improve in China—eventually. China Bus Rev 27(3):22–29

Clark D (2004) Intellectual property litigation in China. China Bus Rev 1(6):25–29

Cohen WM, Nelson RR, Walsh JP (2000) Protecting their intellectual assets: appropriability conditions and why US manufacturing firms patent (or not). NBER Working Paper 77552. Cambridge

Cyr D, Forst P (1991) Human resources management practices in China: a future perspective. Hum Resour Manag 30(2):199–215

D'Iribarne P (1996) The usefulness of an ethnographic approach to the international comparison of organizations. Int Stud Manag Organ 26(4):30–47

David PA (1985) Clio and the economics of qwerty. Am Econ Rev 75:332–337

Dietz MC, Elton JJ (2004) Getting more form intellectual property. McKinsey Q 4:1–3

Dietz MC, Lin SS-T, Yang L (2005) Protecting intellectual property in China. McKinsey Q 3:6–8

Dosi G, Marengoa L, Pasquali C (2006) How much should society fuel the greed of innovators? On the relations between appropriability, opportunities and rates of innovation. Res Policy 35:1110–1121

EC (2004) European Communities—Protection of Trademarks and Geographical Indications for Agricultural Products and Foodstuffs, 26 Aug 2004 (WT/DS 174, WT/DS 290)

Ehrat M (1997) Kompetenzorientierte, analysegestützte Technologiestrategieerarbeitung. Dissertation, Universität St. Gallen, No 1981, St. Gallen

Eisenhardt KM (1989) Building theories from case study research. Acad Manag Rev 14(4):532–550

EIU (2009) A new ranking of the world's most innovative countries. Economist Intelligence Unit

Elizur D, Borg I, Hunt R, Beck IM (1991) The structure of work values: a cross cultural comparison. J Organ Behav 12(1):21–38

Ernst H (1997) The use of patent data for technological forecasting: the diffusion of CNC-technology in the machine tool industry. Small Bus Econ 9:361–381

Ernst H (1998) Industrial research as a source of important patents. Res Policy 27(1):1–15

Ernst H (1999) Führen Patentanmeldungen zu einem nachfolgenden Anstieg des Unternehmenserfolges? Eine Panelanalyse. Z Betriebswirtsch Forsch 51(12):1146–1167

Ernst H (2003) Patent information for strategic technology management. World Pat Inf 25:233–242

Ernst H, Omland N (2003) Patentmanagement junger Technologieunternehmen. Z Betriebswirtsch (2):95–113

EUC (2009) Report on EU Customs Enforcement of Intellectual Property Rights—Results at the EU Border 2009. European Commission, Taxation and Customs Union

Fai FM (2005) Using intellectual property data to analyse China's growing technological capabilities. World Pat Inf 27:49–61

Feng P (2003) Intellectual property in China, 2nd edn. Sweet & Maxwell Asia, Hong Kong

Ferguson N (2011) Civilization: the west and the rest. Penguin, London

Fischer WA, Zedtwitz M von (2004) Chinese R&D: naissance, renaissance, or mirage? R&D Manag 34(4):349

Foley (2009) The Chint V. Schneider settlement: 157 million reasons to believe Chinese patent holder's rights have muscle. Foley & Lardner LLP, Chicago. http://www.foley.com/publications/pub_detail.aspx?pubid=5949

Foray D et al (2002) European Policy for Intellectual Property (EPIP). Working paper of the Institut pour le management de la recherche et de l'innovation (IMRI), Université Paris Dauphine

FTD (2005) Airbus baut 150 Flugzeuge für China. Financial Times Deutschland

Fuchs HJ (2006) Piraten, Fälscher und Kopierer. Gabler, Wiesbaden

Gallini NT (2002) The economics of patents: lessons from recent U.S. patent reform. J Econ Perspect 16(2):131–154

Gassmann O (1999) Praxisnähe mit Fallstudienforschung. Wissenschaftsmanagement 5(3):11–16

Gassmann O, Bader M (2010) Patentmanagement, 3. Aktualisierte Aufl. Springer, Berlin, p 367

Gassmann O, Beckenbauer A (2009) Der Kampf gegen Piraterie ist ein Mehrfrontenkrieg. i.o. N Manag 5:20–23

Gassmann O, Beckenbauer A (2010) Mit den Waffen der Wissensgesellschaft gegen illegale Imitatoren. Innov Manag 7:98–105

Gassmann O, Han Z (2004) Motivations and barriers of foreign R&D activities in China. R&D Manag 34(4):423–437

Gassmann O, Keupp MM (2007) The competitive advantage of early and rapidly internationalising SMEs in the biotechnology industry: a knowledge based view. J World Bus, Special Issue: The Early and Rapid Internationalisation of the Firm 42(3):350–366

Gassmann O, Keupp MM (2008) The Internationalisation of Western firms's R&D in China. Int J Entrep Small Bus 6(4):536–561

Gassmann O, Beckenbauer A, Bader MA (2008) Massnahmen gegen Produktpiraterie am Beispiel Chinas. Innov Manag 2:84–87

GDCh (2008) Gesellschaft Deutscher Chemiker. Chemiestudien-gänge in Deutschland, Statistische Daten 2007. GDCh, Frankfurt a. M.

Gebauer H, Fleisch E, Fischer T (2008) Redefining product strategies in China: overcoming barriers to enter the medium market segment. Strateg Dir 24(5):3–5

Gibbert M et al (2008) What passes as a rigorous case study? Research notes and commentaries. Strateg Manag J 29:1465–1474

Glass AJ, Saggi K (2002) Intellectual property rights and foreign direct investment. J Int Econ 56(2):387–410

Goldberg MD, Feder JM (1991) China's intellectual property legislation. China Bus Rev 18(5):8–11

Goldman Sachs (2004) Growth and Development: The path to 2050

Grant RM (1996) Prospering in dynamically-competitive environments: organizational capability as knowledge integration. Organ Sci 7(4):375–387

Grindley P, Teece D (1997) Managing intellectual capital: licensing and cross-licensing in semiconductors and electronics. Calif Manag Rev 39(2):8–41

Hall DL, Ames RT (1995) Anticipating China: thinking through the narratives of Chinese and Western culture. State University of New York Press, New York

Hammel F (2006) Investment in China. Fortschritte bei den rechtlichen Rahmenbedingungen. VBKI-Magazin 1

Haour G, Von Zedtwitz M (2004) China auf dem Weg zum globalen Innovationslabor. IO N Manag 4:16–20

Harabi N (1995) Appropriability of technical innovations: an empirical analysis. Res Policy 24:981–992

Harhoff D, Reitzig M (2001) Strategien zur Gewinnmaximierung bei der Anmeldung von Patenten. Z Betriebswirtsch 71(5):509

Henley JS, Nyaw MK (1990) The system of management and performance of joint ventures in China: some evidence from Shenzen special economic zone. Adv Chin Ind Stud 1(B):277–295

Hurmelinna-Laukkanen P, Puumalainen K (2007) Nature and dynamics of appropriability: strategies for appropriating returns on innovation. R&D Manag 7(2):95–102

ICC (2006) Current and emerging intellectual property issues for business—a roadmap for business and policy makers. 450/911 Rev. 7. ICC, Paris

ICC (2007) The international anti-counterfeiting directory 2007. International Chamber of Commerce, Counterfeiting Intelligence Bureau. ICC, Barking

Immelt JR, Govindarajan V, Trimble C (2009) How GE is disrupting itself. Harv Bus Rev 87(10):56–65

Jackson T (1998) Foreign companies and Chinese workers: employee motivation in the People's Republic of China. J Organ Chang Manag 11(4):282–300

Keupp MM, Beckenbauer A, Gassmann O (2009) How managers protect intellectual property rights in China using de facto strategies. R&D Manag J 39(2):211–224

Keupp MM, Beckenbauer A, Gassmann O (2010) Enforcing intellectual property rights in weak appropriability regimes: the case of de Facto protection strategies in China. Manag Int Rev 50(1):109–130

Keupp MM, Palmiè M, Gassmann O (2011) Achieving subsidiary integration in international innovation by managerial "Tools". Manag Int Rev 51(2):213–239

Klevorick AK, Levin RC, Nelson R, Winter SG (1995) On the sources and significance of interindustry differences in technological opportunities. Res Policy 24(2):185–205

Kluwer (2005) China intellectual property law guide. Kluwer Law International, Netherlands

Kromrey H (1995) Empirische Sozialforschung: Modelle und Methoden der Datenerhebung und Datenauswertung, 7th edn. Leske & Budrich, Opladen

Kubicek H (1977) Heuristische Bezugsrahmen und heuristisch angelegte Forschungsdesigns als Elemente einer Konstruktionsstrategie empirischer Forschung. In: Köhler R (ed) Empirische und handlungstheoretische Forschungskonzeption in der Betriebswirtschaftslehre. Poeschel, Stuttgart, pp 3–36

La Croix S, Konan D (2002) Intellectual property rights in China: the changing political economy of Chinese-American interests. World Econ 25(6):759–788

Lamnek S (1995) Qualitative Sozialforschung—Methoden und Techniken, 3rd edn. Beltz, Weinheim

Levin SG, Levin SL, Meisel JB (1987) A dynamic analysis of the adoption of a new technology: the case of optical scanners. Rev Econ Stat 69(1):12–17

Li J, Lam K, Sun JJM, Liu SXY (2008) Strategic human resource management, institutionalization, and employment modes: an empirical study in China. Strateg Manag J 29:337–342

Liu JP (2003) The DMCA and the regulation of scientific research. Berkeley Technol Law J 18(2):501–537

Luo Y (1997) Guanxi and performance of foreign-invested enterprises in China: an empirical inquiry. Manag Int Rev 37(1):51–70

Luo Y (2007) Guanxi and business, 2nd edn. World Scientific Publishing, New York, pp 25, 49, 79

Mansfield E (1986) Patents and innovation: an empirical study. Manag Sci 32(2):173–182

Mansfield E, Schwartz M, Wagner S (1981) Imitation costs and patents: an empirical study. Econ J 91(364):907–918

Mazzoleni R, Nelson RR (1998) The benefits and costs of strong patent protection: a contribution to the current debate. Res Policy 27:273–284

McGaughey SL, Liesch PW, Poulson D (2000) An unconventional approach to intellectual property protection: the case of an Australian firm transferring shipbuilding technologies to China. J World Bus 35(1):1–20

McKendrick D (1995) Sources of imitation: improving bank process capabilities. Res Policy 24:783–802

Miles MB, Huberman AM (1994) Qualitative data analysis: an expanded sourcebook, 2nd edn. Sage, London

Misonne B, Ranjard P (2006) Study 12: exploring China's IP environment—strategies and policies. Study on the future opportunities and challenges of EU-China trade and investment relations. http://trade.ec.europa.eu/doclib/docs/2007/february/tradoc_133314.pdf

Mitchel W (1989) Whether and when? Probability and time of incumbents' entry into emerging industrial subfields. Adm Sci Q 34:208–230

Moga TT, Raiti J (2002) The TRIPS agreement and China. China Bus Rev 29(6):12–18

Nevis E (1983) Cultural assumptions and productivity: the United States and China. Sloan Manag Rev 24:17–29

Nonaka I, Takeuchi H (1995) The knowledge creating company: how Japanese companies create the dynamics of innovation. Oxford University Press, New York, p 284

OECD (2005) OECD science, technology and industry: Scoreboard 2005. OECD Rights and Translation Unit, Public Affairs and Communication Directorate, Paris

OECD (2007) The economic impact of counterfeiting and piracy. Secretary-General of the OECD. http://www.oecd.org/dataoecd/13/12/38707619.pdf

OECD (2008) OECD reviews of innovation policy: China. http://www.oecd.org/document/44/0,3746,en_2649_34273_41204780_1_1_1_1,00.html

Oksenberg M, Potter P, Abnett WB (1996) Easing the IPR problem in China's foreign economic relations. National Bureau of Asian Research, Seattle

Park SH, Luo Y (2001) Guanxi and organizational dynamics: organizational networking in Chinese firms. Strateg Manag J 22(5):455–477

Pisano G (2006) Profiting from innovation and the intellectual property revolution. Res Policy 35:1122–1130

Punch KF (1998) Introduction to social research: quantitative and qualitative approaches. Sage, London

Reed R, DeFillippi RJ (1990) Casual ambiguity, barriers to imitation, and sustainable competitive advantage. Acad Manag Rev 15(1):88–103

Reger G (2001) Risikoreduktion durch Technologie-Früherkennung. In: Gassmann O, Kobe C, Voit E (eds) High-Risk-Projekte—Quantensprünge in der Entwicklung erfolgreich managen. Springer, Berlin

Schermerhorn JR, Nyaw M-K (1990) Managerial leadership in Chinese industrial enterprises. Int Stud Manag Organ 20(1–2):9–21

Schlesinger MN (1997) Intellectual property law in China: Part I—Complying with TRIPs requirements. East Asian Exec Rep 19(1):9–20

Schneiderman AM (2007) Filing international patent applications under the Patent Cooperation Treaty (PCT): strategies for delaying costs and maximizing the value of your intellectual property worldwide. In: Krattiger A, Mahoney RT, Nelsen L et al (eds) Intellectual property management in health and agricultural innovation: a handbook of best practices, Chap. 10.7. MIHR, Oxford, and PIPRA, Davis, pp 941–952

Shane S (2001) Technological opportunities and new firm creation. Manag Sci 47(2):205–220

Shapiro C (2001) Navigating the patent thicket: cross licenses, patent pools, and standard setting. In: Jaffe AB, Lerner J, Stern J (eds) Innovation policy and the economy, vol 1. The MIT Press, Cambridge, pp 119–150

SIPO (2007) Annual Report 2007. State Intellectual Property Office of the People's Republic of China

SIPO (2009) Statistics. State intellectual property office of the people's Republic of China. http://www.sipo.gov.cn/sipo_English/statistics/

Sitte B (2006) Schutzmaßnahmen gegen chinesische Produkt- und Markenpiraterie. Diplomica, Hamburg

Smith M, Hansen F (2002) Managing IP: a strategic point of view. J Intellect Cap 3(4):366–374

Staake TR, Fleisch E (2007) Countering counterfeit trade. Springer, Berlin

Stake RE (1995) The art of case study research. Sage, Thousand Oaks

Sullivan P (2001) Profiting from intellectual capital. Wiley, New York

Sun Y (2003) Determinants of foreign patents in China. World Pat Inf 25(1):27–37

Teece DJ (1986) Profiting from technological innovation: implications for integration, collaboration, licensing and public policy. Res Policy 15(6):285–305

Teece DJ (2006) Reflections on 'Profiting from Innovation'. Res Policy 35:1131–1146

The Word Factbook (2006) The World Factbook. Central Intelligence Agency. https://www.cia.gov/cia/publications/factbook/index.html

Tom G, Garibaldi B, Zeng Y, Pilcher J (1998) Consumer demand for counterfeit goods. Psychol Mark 15(5):405–421

Tomczak T (1992) Forschungsmethoden in der Marketingwissenschaft. Ein Plädoyer für den qualitativen Forschungsansatz. Mark ZFP 2:28–37

Tung RL (1991) Motivation in Chinese industrial enterprises. In: Steers RM, Porter LW (eds) Innovation and work behavior, 5th edn. McGraw Hill, New York

Tymon WG, Stumpf SA, Doh JP (2010) Exploring talent management in India: the neglected role of intrinsic rewards. J World Bus 45:109–121

Ulrich H (1981) Die Betriebswirtschaftslehre als anwendungsorientierte Sozialwissenschaft. In: Geist MN, Köhler R (eds) Die Führung des Betriebes. Poeschel, Stuttgart, pp 1–26

Ulrich H, Krieg W (1974) St. Galler management-modell, 3rd edn. Paul Haupt, Bern

USE (2010) Issues in focus. Intellectual property rights. Patent. U.S. Department of State, Embassy of the United States, Beijing. http://beijing.usembassy-china.org.cn/iprpatent.html

Utterback J, Suarez F (1993) Innovation, competition, and market structure. Res Policy 22(1):1–21

Van de Ven AH (2006) Engaged Scholarship: ceating knowledge for science and practice. University of Minnesota, Mineapolis

Von Welser M, González A (2007) Marken- und Produktpiraterie. Strategien und Lösungsansätze zu ihrer Bekämpfung. Wiley, Weinheim

Von Zedtwitz M (1999) Managing interfaces in international R&D. Difo-Druck, Bamberg

Von Zedtwitz M (2004) Managing foreign R&D laboratories in China. R&D Manag 34(4):439

Von Zedtwitz M, Birkinshaw J, Gassmann O (eds) (2008) International management of research and development. Edward Elgar, Massachusetts, p 579

Voss C, Tsikriktsis N, Frohlich M (2002) Case research in operation management. Int J Oper Product Manag 22(2):196–219

Walder AG (1986) Communist Neo-traditionalism. University of California Press, Berkeley

Weinstein V, Fernandez D (2004) Recent developments in China's intellectual property laws. Chin J Int Law 3(1):227–240

Weldon E, Vanhonacker W (1999) Operating a foreign-invested enterprise in China: challenges for managers and management researchers. J World Bus 34(1):94–107

Wessel VW (1993) Technology transfer and intellectual property: an analysis of the NASA approach. Technovation 13(3):133–146

WHO (2011) World Health Organization: general information on counterfeit medicines. http://www.who.int/medicines/services/counterfeit/overview/en/

Winter SG (2006) The logic of appropriability: from Schumpeter to arrow to Teece. Res Policy 35(8):1100–1106

Yang D (2003a) Intellectual property and doing business in China. Pergamon, London

Yang D (2003b) The development of intellectual property in China. World Pat Inf 25:131–142

Yang D (2005) Culture matters to multinationals' intellectual property business. J World Bus 40:281–301

Yang D, Clarke P (2005) Globalisation and intellectual property in China. Technovation 25:545–555

Yang D, Fryxell GE, Sie AKY (2008) Anti-piracy effectiveness and managerial confidence: insights from multinationals in China. J World Bus 43:321–339

Yin RK (1994) Case study research: design and methods. Sage, Thousand Oaks

Yin RK (2003) Applications of case study research, 2nd edn. Sage, Thousand Oaks

Young L (2005) China grants intellectual property protection for Syngenta herbicide ingredient. Chem Week 167(9):15–30

Zeng M, Williamson PJ (2007) Dragons at your door: how Chinese cost innovation is disrupting global competition. Harvward Business School Press, Boston

Zeschky M, Widenmayer B, Gassmann O (2010) How Mettler Toledo organizes for value innovation. Res Technol Manag, Special Issue: Innovation in Emerging Markets

Zhang C, Zeng DZ, Mako WP, Seward J (2009) Promoting enterprise-led innovation in China (Directions in Development) (Directions in Development, Countries and Regions). The World Bank, Washington, DC

Ziedonis RH (2004) Don't fence me. Fragmented markets for technology and the patent acquisition strategies of firms. Manag Sci 50(6):804–820

Index